I0030976

Basic Machines & How They Work

The 1994 Edition
Complete with
Study Materials

Naval Education and Training Program
Development Center

STONE BASIN
BOOKS

Although the words "he," "him," and "his" are used sparingly in this course to enhance communication, they are not intended to be gender driven or to affront or discriminate against anyone.

Published in 2016 by Stone Basin Books

Copyright © 1994 U.S. Department of the Navy

Basic Machines and How They Work
ISBN: 978-1-62654-585-4 (paperback)
978-1-62654-586-1 (casebound)
978-1-62654-587-8 (spiralbound)

Cover design by Adrienne Núñez
Editorial and proofreading assistance by Ian Straus,
Echo Point Books & Media

PREFACE

By enrolling in this self-study course, you have demonstrated a desire to improve yourself and the Navy. Remember, however, this self-study course is only one part of the total Navy training program. Practical experience, schools, selected reading, and your desire to succeed are also necessary to successfully round out a fully meaningful training program.

COURSE OVERVIEW: In completing this nonresident training course, you will demonstrate a knowledge of the subject matter by correctly answering questions on the following: concepts and principles of operation of basic mechanical devices, and the construction and method of operation of common mechanical devices, such as engines and transmissions.

THE COURSE: This self-study course is organized into subject matter areas, each containing learning objectives to help you determine what you should learn along with text and illustrations to help you understand the information. The subject matter reflects day-to-day requirements and experiences of personnel in the rating or skill area. It also reflects guidance provided by Enlisted Community Managers (ECMs) and other senior personnel, technical references, instructions, etc., and either the occupational or naval standards, which are listed in the *Manual of Navy Enlisted Manpower Personnel Classifications and Occupational Standards*, NAVPERS 18068.

THE QUESTIONS: The questions that appear in this course are designed to help you understand the material in the text.

VALUE: In completing this course, you will improve your military and professional knowledge. Importantly, it can also help you study for the Navy-wide advancement in rate examination. If you are studying and discover a reference in the text to another publication for further information, look it up.

1994 Edition Prepared by
AMHC(AW) Edward L. Prater

Published by
NAVAL EDUCATION AND TRAINING
PROFESSIONAL DEVELOPMENT
AND TECHNOLOGY CENTER

NAVSUP Logistics Tracking Number
0504-LP-026-7140

Sailor's Creed

"I am a United States Sailor.

I will support and defend the Constitution of the United States of America and I will obey the orders of those appointed over me.

I represent the fighting spirit of the Navy and those who have gone before me to defend freedom and democracy around the world.

I proudly serve my country's Navy combat team with honor, courage and commitment.

I am committed to excellence and the fair treatment of all."

CONTENTS

INSTRUCTIONS FOR TAKING THE COURSE

ASSIGNMENTS

The text pages that you are to study are listed at the beginning of each assignment. Study these pages carefully before attempting to answer the questions. Pay close attention to tables and illustrations and read the learning objectives. The learning objectives state what you should be able to do after studying the material. Answering the questions correctly helps you accomplish the objectives.

SELECTING YOUR ANSWERS

Read each question carefully, then select the BEST answer. You may refer freely to the text. The answers must be the result of your own work and decisions. You are prohibited from referring to or copying the answers of others and from giving answers to anyone else taking the course.

SUBMITTING YOUR ASSIGNMENTS

To have your assignments graded, you must be enrolled in the course with the Nonresident Training Course Administration Branch at the Naval Education and Training Professional Development and Technology Center (NETPDTC). Following enrollment, there are two ways of having your assignments graded: (1) use the Internet to submit your assignments as you complete them, or (2) send all the assignments at one time by mail to NETPDTC.

Grading on the Internet: Advantages to Internet grading are:

- you may submit your answers as soon as you complete an assignment, and
- you get your results faster; usually by the next working day (approximately 24 hours).

In addition to receiving grade results for each assignment, you will receive course completion confirmation once you have completed all the assignments. To submit your assignment answers via the Internet, go to:

http://courses.cnet.navy.mil

Grading by Mail: When you submit answer sheets by mail, send all of your assignments at one time. Do NOT submit individual answer sheets for grading. Mail all of your assignments in an envelope, which you either provide yourself or obtain from your nearest Educational Services Officer (ESO). Submit answer sheets to:

> COMMANDING OFFICER
> NETPDTC N331
> 6490 SAUFLEY FIELD ROAD
> PENSACOLA FL 32559-5000

Answer Sheets: All courses include one "scannable" answer sheet for each assignment. These answer sheets are preprinted with your SSN, name, assignment number, and course number. Explanations for completing the answer sheets are on the answer sheet.

Do not use answer sheet reproductions: Use only the original answer sheets that we provide—reproductions will not work with our scanning equipment and cannot be processed.

Follow the instructions for marking your answers on the answer sheet. Be sure that blocks 1, 2, and 3 are filled in correctly. This information is necessary for your course to be properly processed and for you to receive credit for your work.

COMPLETION TIME

Courses must be completed within 12 months from the date of enrollment. This includes time required to resubmit failed assignments.

PASS/FAIL ASSIGNMENT PROCEDURES

If your overall course score is 3.2 or higher, you will pass the course and will not be required to resubmit assignments. Once your assignments have been graded you will receive course completion confirmation.

If you receive less than a 3.2 on any assignment and your overall course score is below 3.2, you will be given the opportunity to resubmit failed assignments. **You may resubmit failed assignments only once.** Internet students will receive notification when they have failed an assignment--they may then resubmit failed assignments on the web site. Internet students may view and print results for failed assignments from the web site. Students who submit by mail will receive a failing result letter and a new answer sheet for resubmission of each failed assignment.

COMPLETION CONFIRMATION

After successfully completing this course, you will receive a letter of completion.

ERRATA

Errata are used to correct minor errors or delete obsolete information in a course. Errata may also be used to provide instructions to the student. If a course has an errata, it will be included as the first page(s) after the front cover. Errata for all courses can be accessed and viewed/downloaded at:

http://www.advancement.cnet.navy.mil

STUDENT FEEDBACK QUESTIONS

We value your suggestions, questions, and criticisms on our courses. If you would like to communicate with us regarding this course, we encourage you, if possible, to use e-mail. If you write or fax, please use a copy of the Student Comment form that follows this page.

For subject matter questions:

E-mail: n314.products@cnet.navy.mil
Phone: Comm: (850) 452-1001, Ext. 1826
           DSN: 922-1001, Ext. 1826
           FAX: (850) 452-1370
           (Do not fax answer sheets.)
Address: COMMANDING OFFICER
           NETPDTC N314
           6490 SAUFLEY FIELD ROAD
           PENSACOLA FL 32509-5237

For enrollment, shipping, grading, or completion letter questions

E-mail: fleetservices@cnet.navy.mil
Phone: Toll Free: 877-264-8583
           Comm: (850) 452-1511/1181/1859
           DSN: 922-1511/1181/1859
           FAX: (850) 452-1370
           (Do not fax answer sheets.)
Address: COMMANDING OFFICER
           NETPDTC N331
           6490 SAUFLEY FIELD ROAD
           PENSACOLA FL 32559-5000

NAVAL RESERVE RETIREMENT CREDIT

If you are a member of the Naval Reserve, you may earn retirement points for successfully completing this course, if authorized under current directives governing retirement of Naval Reserve personnel. For Naval Reserve retirement, this course is evaluated at 6 points. (Refer to *Administrative Procedures for Naval Reservists on Inactive Duty,* BUPERSINST 1001.39, for more information about retirement points.)

<u>Student Comments</u>

Course Title: *Basic Machines* _____

NAVEDTRA: 14037 _____ **Date**: _____

<u>We need some information about you</u>:

Rate/Rank and Name: _____ SSN: _____ Command/Unit _____

Street Address: _____ City: _____ State/FPO: _____ Zip _____

<u>Your comments, suggestions, etc</u>.:

Privacy Act Statement: Under authority of Title 5, USC 301, information regarding your military status is requested in processing your comments and in preparing a reply. This information will not be divulged without written authorization to anyone other than those within DOD for official use in determining performance.

NETPDTC 1550/41 (Rev 4-00

CHAPTER 1

LEVERS

CHAPTER LEARNING OBJECTIVES

Upon completion of this chapter, you should be able to do the following:

- *Explain the use of levers when operating machines afloat and ashore.*

- *Discuss the classes of levers.*

Through the ages, ships have evolved from crude rafts to the huge complex cruisers and carriers of today's Navy. It was a long step from oars to sails, another long step from sails to steam, and another long step to today's nuclear power. Each step in the progress of shipbuilding has involved the use of more and more machines.

Today's Navy personnel are specialists in operating and maintaining machinery. Boatswains operate winches to hoist cargo and the anchor; personnel in the engine room operate pumps, valves, generators, and other machines to produce and control the ship's power; personnel in the weapons department operate shell hoists and rammers and elevate and train the guns and missile launchers; the cooks operate mixers and can openers; personnel in the CB ratings drive trucks and operate cranes, graders, and bulldozers. In fact, every rating in the Navy uses machinery sometime during the day's work.

Each machine used aboard ship has made the physical work load of the crew lighter; you don't walk the capstan to raise the anchor, or heave on a line to sling cargo aboard. Machines are your friends. They have taken much of the backache and drudgery out of a sailor's lift. Reading this book should help you recognize and understand the operation of many of the machines you see about you.

WHAT IS A MACHINE?

As you look about you, you probably see half a dozen machines that you don't recognize as such. Ordinarily you think of a machine as a complex device-a gasoline engine or a typewriter. They are machines; but so are a hammer, a screwdriver, a ship's wheel. A machine is any device that helps you to do work. It may help by changing the amount of force or the speed of action. A claw hammer, for example, is a machine. You can use it to apply a large force for pulling out a nail; a relatively small pull on the handle produces a much greater force at the claws.

We use machines to *transform* energy. For example, a generator transforms mechanical energy into electrical energy. We use machines to *transfer* energy from one place to another. For example, the connecting rods, crankshaft, drive shaft, and rear axle of an automobile transfer energy from the engine to the rear wheels.

Another use of machines is to multiply force. We use a system of pulleys (a chain hoist, for example) to lift a heavy load. The pulley system enables us to raise the load by exerting a force that is smaller than the weight of the load. We must exert this force over a greater distance than the height through which the load is raised; thus, the load will move slower than the chain on which we pull. The machine enables us to gain force, but only at the expense of speed.

Machines may also be used to multiply speed. The best example of this is the bicycle, by which we gain speed by exerting a greater force.

Machines are also used to change the direction of a force. For example, the Signalman's halyard enables one end of the line to exert an upward force on a signal flag while a downward force is exerted on the other end.

There are only six simple machines: the lever, the block, the wheel and axle, the inclined plane, the screw, and the gear. Physicists, however, recognize only two basic principles in machines: those of the lever and the inclined plane. The wheel and axle, block and tackle, and gears may be considered levers. The wedge and the screw use the principle of the inclined plane.

When you are familiar with the principles of these simple machines, you can readily understand the

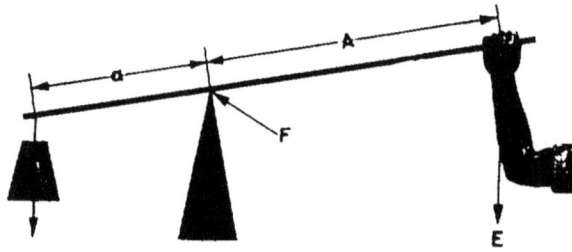

Figure 1-1.-A simple lever.

operation of complex machines. Complex machines are merely combinations of two or more simple machines.

THE LEVER

The simplest machine, and perhaps the one with which you are most familiar, is the lever. A seesaw is a familiar example of a lever in which one weight balances the other.

You will find that all levers have three basic parts: the fulcrum (F), a force or effort (E), and a resistance (R). Look at the lever in figure 1-1. You see the pivotal point (fulcrum) (F); the effort (E), which is applied at a distance (A) from the fulcrum; and a resistance (R), which acts at a distance (a) from the fulcrum. Distances A and a are the arms of the lever.

CLASSES OF LEVERS

The three classes of levers are shown in figure 1-2. The location of the fulcrum (the fixed or pivot point) in relation to the resistance (or weight) and the effort determines the lever class.

First Class

In the first class (fig. 1-2, part A), the fulcrum is located between the effort and the resistance. As mentioned earlier, the seesaw is a good example of a first-class lever. The amount of weight and the distance from the fulcrum can be varied to suit the need.

Notice that the sailor in figure 1-3 applies effort on the handles of the oars. An oar is another good example. The oarlock is the fulcrum, and the water is the resistance. In this case, as in figure 1-1, the force is applied on one side of the fulcrum and the resistance to be overcome is applied to the opposite side; hence, this is a first class lever. Crowbars, shears, and pliers are common examples of this class of levers.

Second Class

The second class of lever (fig. 1-2, part B) has the fulcrum at one end, the effort applied at the other end, and the resistance somewhere between those points. The

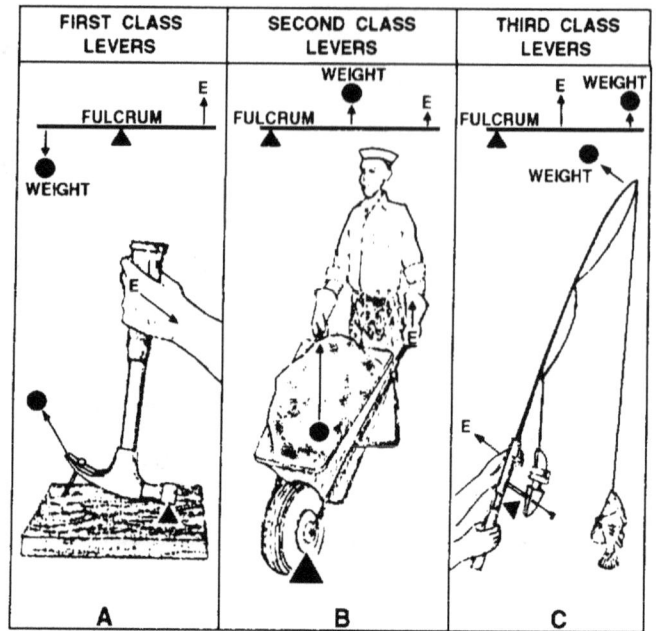

Figure 1-2.-Three classes of levers.

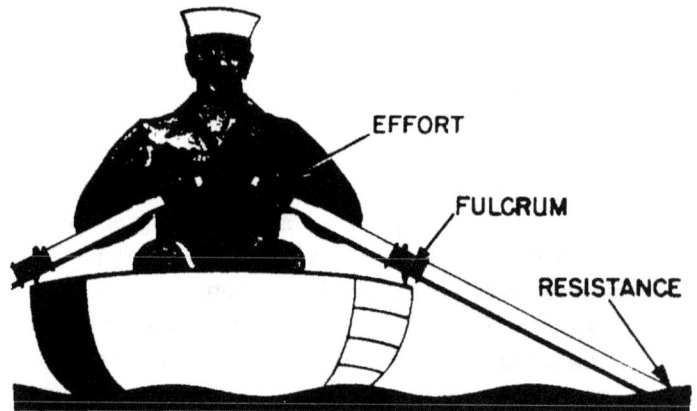

Figure 1-3.-Oars are levers.

wheelbarrow in figure 1-4 is a good example of a second-class lever. If you apply 50 pounds of effort to the handles of a wheelbarrow 4 feet from the fulcrum (wheel), you can lift 200 pounds of weight 1 foot from the fulcrum. If the load were placed farther away from the wheel, would it be easier or harder to lift?

Levers of the first and second class are commonly used to help in overcoming big resistances with a relatively small effort.

Third Class

Sometimes you will want to speed up the movement of the resistance even though you have to use a large amount of effort. Levers that help you accomplish this are in the third class of levers. As shown in figure 1-2, part C, the fulcrum is at one end of the lever, and the

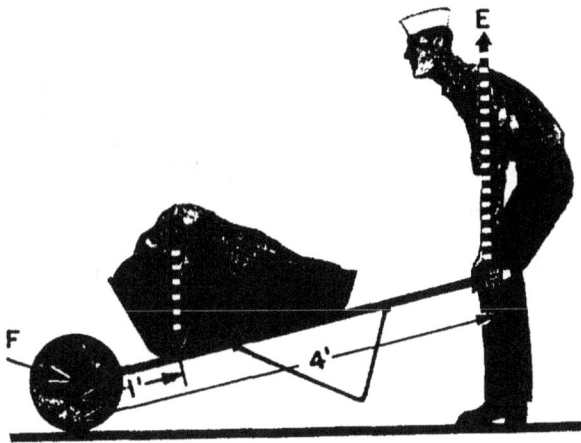

Figure 1-4.-This makes it easier.

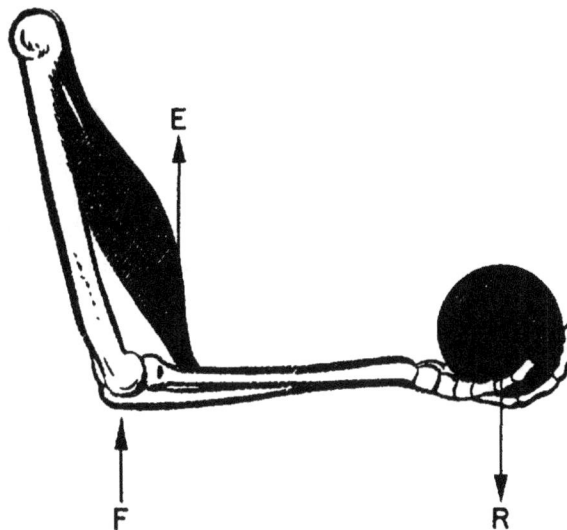

Figure 1-6.-Your arm is a lever.

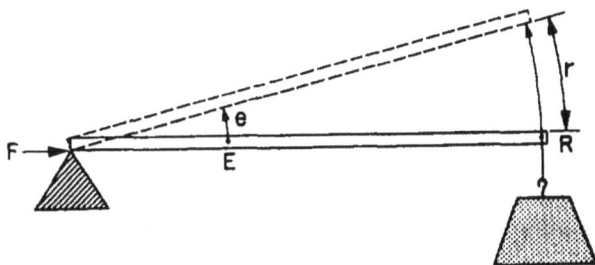

Figure 1-5.-A third-class lever.

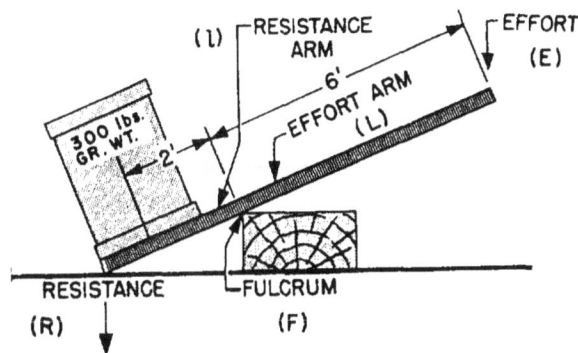

Figure 1-7.-Easy does it.

weight or resistance to be overcome is at the other end, with the effort applied at some point between. You can always spot the third-class levers because you will find the effort applied between the fulcrum and the resistance. Look at figure 1-5. It is easy to see that while E moved the short distance (e), the resistance (R) was moved a greater distance (r). The speed of R must have been greater than that of E, since R covered a greater distance in the same length of time.

Your arm (fig. 1-6) is a third-class lever. It is this lever action that makes it possible for you to flex your arms so quickly. Your elbow is the fulcrum. Your biceps muscle, which ties onto your forearm about an inch below the elbow, applies the effort; your hand is the resistance, located about 18 inches from the fulcrum. In the split second it takes your biceps muscle to contract an inch, your hand has moved through an 18-inch arc. You know from experience that it takes a big pull at E to overcome a relatively small resistance at R. Just to experience this principle, try closing a door by pushing on it about 3 or 4 inches from the hinges (fulcrum). The moral is, you don't use third-class levers to do heavy jobs; you use them to gain speed.

One convenience of machines is that you can determine in advance the forces required for their operation, as well as the forces they will exert. Consider for a moment the first class of levers. Suppose you have an iron bar, like the one shown in figure 1-7. This bar is 9 feet long, and you want to use it to raise a 300-pound crate off the deck while you slide a dolly under the crate; but you can exert only 100 pounds to lift the crate. So, you place the fulcrum-a wooden block-beneath one end of the bar and force that end of the bar under the crate. Then, you push down on the other end of the bar. After a few adjustments of the position of the fulcrum, you will find that your 100-pound force will just fit the crate when the fulcrum is 2 feet from the center of the crate. That leaves a 6-foot length of bar from the fulcrum to the point where you pushdown. The 6-foot portion is three times as long as the distance from the fulcrum to the center of the crate. And you lifted a load three times as great as the force you applied (3 x 100 = 300 pounds).

Here is a sign of a direct relationship between the *length of the lever arm and the force acting on that arm.*

You can state this relationship in general terms by saying: the length of the effort arm is the same number of times greater than the length of the resistance arm as the resistance to be overcome is greater than the effort you must apply. Writing these words as a mathematical equation, we have

$$\frac{L}{l} = \frac{R}{E},$$

where

L = length of effort arm,

l = length of resistance arm,

R = resistance weight or force, and

E = effort force.

Remember that all distances must be in the same units, such as feet, and that all forces must be in the same units, such as pounds.

Now let's take another problem and see how it works out. Suppose you want to pry up the lid of a paint can (fig. 1-8) with a 6-inch file scraper, and you know that the average force holding the lid is 50 pounds. If the distance from the edge of the paint can to the edge of the cover is 1 inch, what force will you have to apply on the end of the file scraper?

According to the formula,

$$\frac{L}{l} = \frac{R}{E},$$

here,

L = 5 inches

l = 1 inch

R = 50 pounds, and

E is unknown.

Then, substituting the numbers in their proper places, we have

$$\frac{5}{l} = \frac{50}{E}$$

and

$$E = \frac{50 \times 1}{5} = 10 \text{ pounds}$$

You will need to apply a force of only 10 pounds.

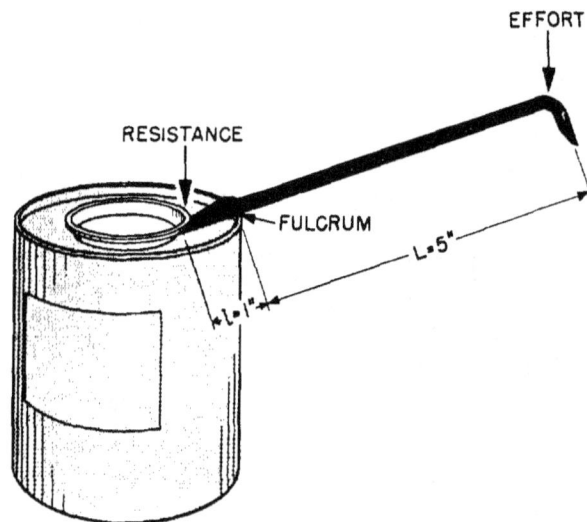

Figure 1-8.-A first-class job.

The same general formula applies for the second class of levers; but you must be careful to measure the proper lengths of the effort arm and the resistance arm. Looking back at the wheelbarrow problem, assume that the length of the handles from the axle of the wheel—which is the fulcrum-to the grip is 4 feet. How long is the effort arm? You're right, it's 4 feet. If the center of the load of sand is 1 foot from the axle, then the length of the resistance arm is 1 foot.

By substituting in the formula,

$$\frac{L}{l} = \frac{R}{E},$$

$$\frac{4}{l} = \frac{200}{E},$$

and

E = 50 pounds.

Now for the third-class lever. With one hand, you lift a projectile weighing approximately 10 pounds. If your biceps muscle attaches to your forearm 1 inch below your elbow and the distance from the elbow to the palm of your hand is 18 inches, what pull must your muscle exert to hold the projectile and flex your arm at the elbow?

By substituting in the formula,

$$\frac{L}{l} = \frac{R}{E},$$

it becomes

$$\frac{l}{18} = \frac{10}{E}.$$

and

$$E = 18 \times 10 = 180 \text{ pounds.}$$

Your muscle must exert a 180-pound pull to hold up a 10-pound projectile. Our muscles are poorly arranged for lifting or pulling-and that's why some work seems pretty tough. But remember, third-class levers are used primarily to speed up the motion of the resistance.

Curved Lever Arms

Up to this point, you have been looking at levers with straight arms. In every case, the direction in which the resistance acts is parallel to the direction in which the effort is exerted. However, not all levers are straight. You'll need to learn to recognize all types of levers and to understand their operation.

Look at figure 1-9. You may wonder how to measure the length of the effort arm, which is represented by the curved pump handle. You do not measure around the curve; you still use a straight-line distance. To determine the length of the effort arm, draw a straight line (AB) through the point where the effort is applied and in the direction that it is applied. From point E on this line, draw a second line (EF) that passes through the fulcrum and is perpendicular to line AB. The length of line EF is the actual length (L) of the effort arm.

To find the length of the resistance arm, use the same method. Draw a line (MN) in the direction that the resistance is operating and through the point where the resistance is attached to the other end of the handle. From point R on this line, draw a line (RF) perpendicular to MN so that it passes through the fulcrum. The length of RF is the length (l) of the resistance arm.

Regardless of the curvature of the handle, this method can be used to find lengths L and l. Then, curved levers are solved just like straight levers.

MECHANICAL ADVANTAGE

There is another thing about the first and second classes of levers that you have probably noticed by now. Since they can be used to magnify the applied force, they provide positive mechanical advantages. The third-class lever provides what is called a fractional mechanical advantage, which is really a mechanical disadvantage—you use more force than the force of the load you lift.

In the wheelbarrow problem, you saw that a 50-pound pull actually overcame the 200-pound weight

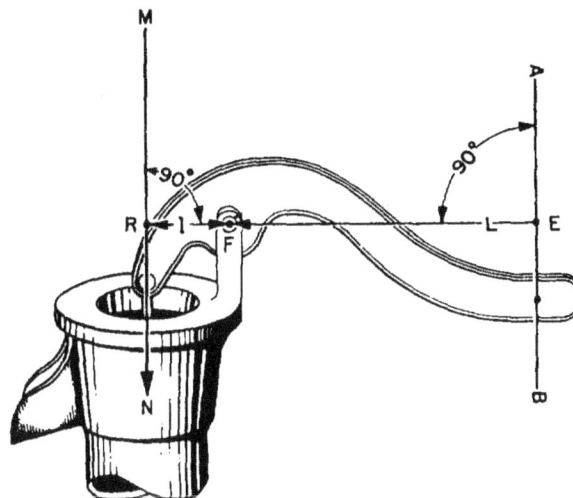

Figure 1-9.-A curved lever arm.

of the sand. The sailor's effort was magnified four times, so you may say that the mechanical advantage of the wheelbarrow is 4. Expressing the same idea in mathematical terms,

$$\text{MECHANICAL ADVANTAGE} = \frac{\text{RESISTANCE}}{\text{EFFORT}}$$

or

$$\text{M.A.} = \frac{R}{E}$$

Thus, in the case of the wheelbarrow,

$$\text{M.A.} = \frac{200}{50} = 4$$

This rule—mechanical advantage equals resistance divided by effort —applies to all machines.

The mechanical advantage of a lever may also be found by dividing the length of effort arm A by the length of resistance arm a. Stated as a formula, this reads:

$$\text{MECHANICAL ADVANTAGE} = \frac{\text{EFFORT ARM}}{\text{RESISTANCE ARM}}$$

or

$$\text{M.A.} = \frac{A}{a}$$

How does this apply to third-class levers? Your muscle pulls with a force of 1,800 pounds to lift a 100-pound projectile. So you have a mechanical advantage of

$$\frac{100}{1,800}, \text{ or } \frac{1}{18},$$

which is fractional-less than 1.

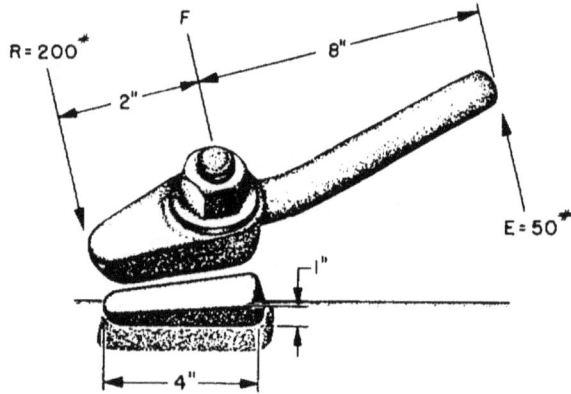

Figure 1-10.-It's a dog.

Figure 1-12.-Using a wrecking bar.

APPLICATIONS AFLOAT AND ASHORE

Doors, called hatches aboard a ship, are locked shut by lugs called dogs. Figure 1-10 shows you how these dogs are used to secure the door. If the handle is four times as long as the lug, that 50-pound heave of yours is multiplied to 200 pounds against the slanting face of the wedge. Incidentally, take a look at the wedge—it's an inclined plane, and it multiplies the 200-pound force by about 4. Result: Your 50-pound heave actually ends up as a 800-pound force on each wedge to keep the hatch closed! The hatch dog is one use of a first-class lever in combination with an inclined plane.

The breech of a big gun is closed with a breech plug. Figure 1-11 shows you that this plug has some interrupted screw threads on it, which fit into similar interrupted threads in the breech. Turning the plug part way around locks it into the breech. The plug is locked and unlocked by the operating lever. Notice that the connecting rod is secured to the operating lever a few inches from the fulcrum. You'll see that this is an application of a second-class lever.

You know that the plug is in there good and tight. But, with a mechanical advantage of 10, your 100-pound pull on the handle will twist the plug loose with a force of a half ton.

If you've spent any time opening crates at a base, you've already used a wrecking bar. The sailor in figure 1-12 is busily engaged in tearing that crate open.

Figure 1-12.-Using a wrecking bar.

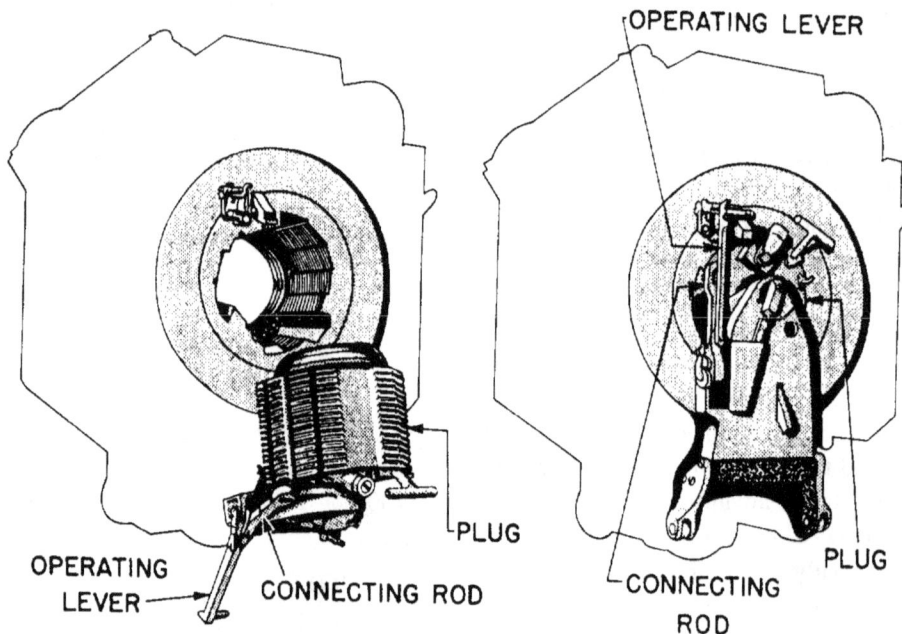

Figure 1-11.-The breech of an 8-inch gun.

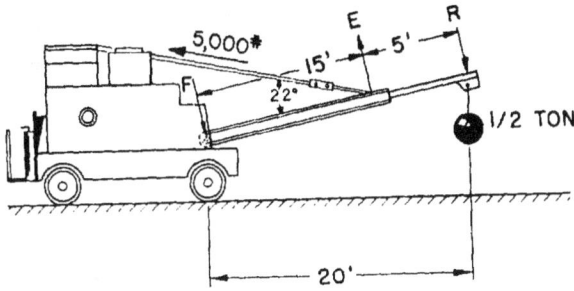

Figure 1-13.-An electric crane.

Figure 1-14.-A. A pelican hook; B. A chain stopper.

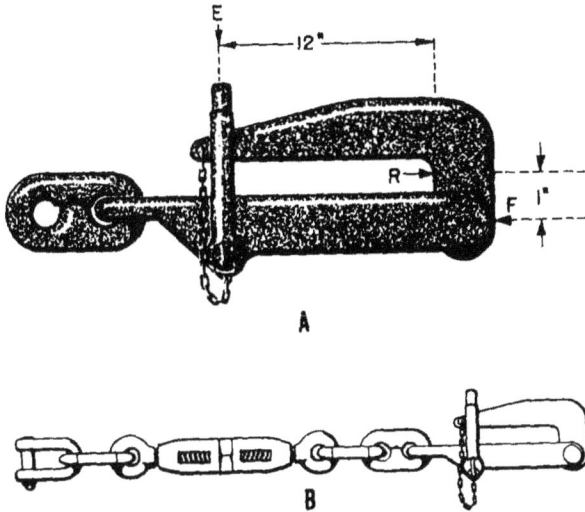

Figure 1-15.-An improvised drill press.

The wrecking bar is a first-class lever. Notice that it has curved lever arms. Can you figure the mechanical advantage of this one? Your answer should be M.A. = 5.

The crane in figure 1-13 is used for handling relatively light loads around a warehouse or a dock. You can see that the crane is rigged as a third-class lever; the effort is applied between the fulcrum and the load. This gives a mechanical advantage of less than 1. If it's going to support that 1/2-ton load, you know that the pull on the lifting cable will have to be considerably greater than 1,000 pounds. How much greater? Use the formula to figure it out:

$$\frac{L}{l} = \frac{R}{E}$$

Got the answer? Right. . . $E = 1,333$ pounds

Now, because the cable is pulling at an angle of about 22° at E, you can use some trigonometry to find that the pull on the cable will be about 3,560 pounds to lift the 1/2-ton weight! However, since the loads are generally light, and speed is important, the crane is a practical and useful machine.

Anchors are usually housed in the hawsepipe and secured by a chain stopper. The chain stopper consists of a short length of chain containing a turnbuckle and a pelican hook. When you secure one end of the stopper to a pad eye in the deck and lock the pelican hook over the anchor chain, the winch is relieved of the strain.

Figure 1-14, part A, gives you the details of the pelican hook.

Figure 1-14, part B, shows the chain stopper as a whole. Notice that the load is applied close to the fulcrum. The resistance arm is very short. The bale shackle, which holds the hook secure, exerts its force at a considerable distance from the fulcrum. If the chain rests against the hook 1 inch from the fulcrum and the bale shackle is holding the hook closed 12 + 1 = 13 inches from the fulcrum, what's the mechanical advantage? It's 13. A strain of only 1,000 pounds on the base shackle can hold the hook closed when a 6 1/2-ton anchor is dangling over the ship's side. You'll recognize the pelican hook as a second-class lever with curved arms.

Figure 1-15 shows you a couple of guys who are using their heads to spare their muscles. Rather than exert themselves by bearing down on that drill, they pick up a board from a nearby crate and use it as a second-class lever.

If the drill is placed halfway along the board, they will get a mechanical advantage of 2. How would you increase the mechanical advantage if you were using this rig? Right. You would move the drill in closer to the fulcrum. In the Navy, a knowledge of levers and how to apply them pays off.

SUMMARY

Now for a brief summary of levers:

Levers are machines because they help you to do your work. They help by changing the size, direction, or speed of the force you apply.

There are three classes of levers. They differ primarily in the relative points where effort is applied, where the resistance is overcome, and where the fulcrum is located.

First-class levers have the effort and the resistance on opposite sides of the fulcrum, and effort and resistance move in opposite directions.

Second-class levers have the effort and the resistance on the same side of the fulrum but the effort is farther from the fulcrum than is the resistance. Both effort and resistance move in the same direction.

Third-class levers have the effort applied on the same side of the fulcrum as the resistance but the effort is applied between the resistance and the fulcrum, and both effort and resistance move in the same direction.

First- and second-class levers magnify the amount of effort exerted and decrease the speed of effort. First-class and third-class levers magnify the distance and the speed of the effort exerted and decrease its magnitude.

The same general formula applies to all three types of levers:

$$\frac{L}{l} = \frac{R}{E}$$

Mechanical advantage (M.A.) is an expression of the ratio of the applied force and the resistance. It may be written:

$$\text{M.A.} = \frac{R}{E}$$

CHAPTER 2

BLOCK AND TACKLE

CHAPTER LEARNING OBJECTIVES

Upon completion of this chapter, you should be able to do the following:

● *Describe the advantage of block and tackle afloat and ashore*

Blocks—pulleys to a landlubber—are simple machines that have many uses aboard ship, as well as onshore. Remember how your mouth hung open as you watched movers taking a piano out of a fourth story window? The guy on the end of the tackle eased the piano safely to the sidewalk with a mysterious arrangement of blocks and ropes. Or, you've been in the country and watched the farmer use a block and tackle to put hay in a barn. Since old Dobbin or the tractor did the hauling, there was no need for a fancy arrangement of ropes and blocks. Incidentally, you'll often hear the rope or tackle called the fall, block and tack, or block and fall.

In the Navy you'll rig a block and tackle to make some of your work easier. Learn the names of the parts of a block. Figure 2-1 will give you a good start on this. Look at the single block and see some of the ways you can use it. If you lash a single block to a fixed object-an overhead, a yardarm, or a bulkhead-you give yourself the advantage of being able to pull from a convenient direction. For example, in figure 2-2 you haul up a flag hoist, but you really pull down. You can do this by having a single sheaved block made fast to the yardarm. This makes it possible for you to stand in a convenient place near the flag bag and do the job. Otherwise you would have to go aloft, dragging the flag hoist behind you.

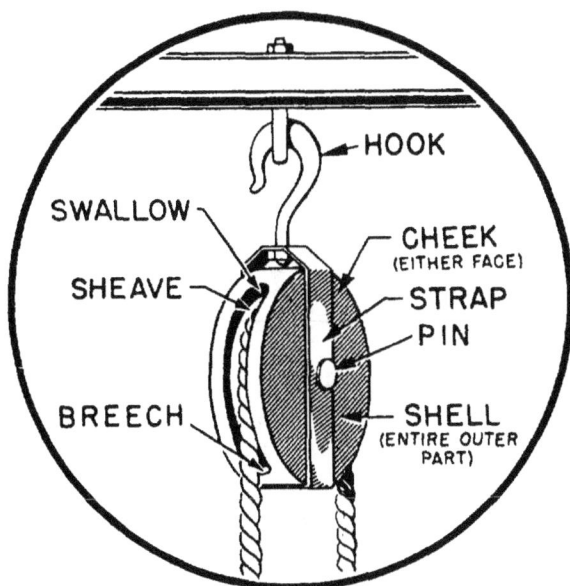

Figure 2-1.-Look it over.

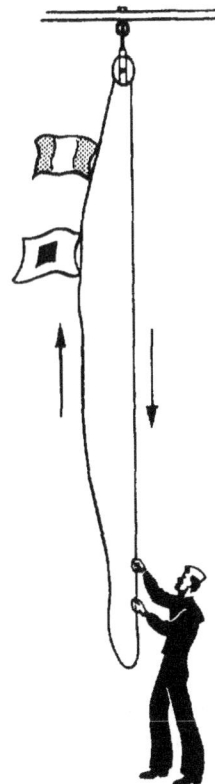

Figure 2-2.-A flag hoist.

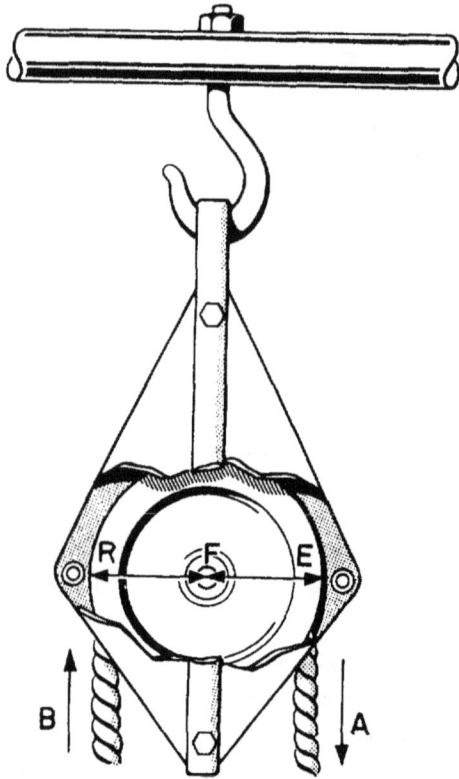

Figure 2-3.-No advantage.

Figure 2-4.-A runner.

MECHANICAL ADVANTAGE

With a single fixed sheave, the force of your down pull on the fall must be equal to the weight of the object hoist. You can't use this rig to lift a heavy load or resistance with a small effort-you can change only the direction of your pull.

A single fixed block is a first-class lever with equal arms. The arms (EF and FR) in figure 2-3 are equal; hence, the mechanical advantage is 1. When you pull down at A with a force of 1 pound, you raise a load of 1 pound at B. A single fixed block does not magnify force nor speed.

You can, however, use a single block and fall to magnify the force you exert. Notice in figure 2-4 that the block is not fixed. The fall is doubled as it supports the 200-pound cask. When rigged this way, you call the single block and fall a runner. Each half of the fall carries one-half of the total bad, or 100 pounds. Thus, with the runner, the sailor is lifting a 200-pound cask with a 100-pound pull. The mechanical advantage is 2. Check this by the formula:

$$\text{M.A.} = \frac{R}{E} = \frac{200}{100}$$

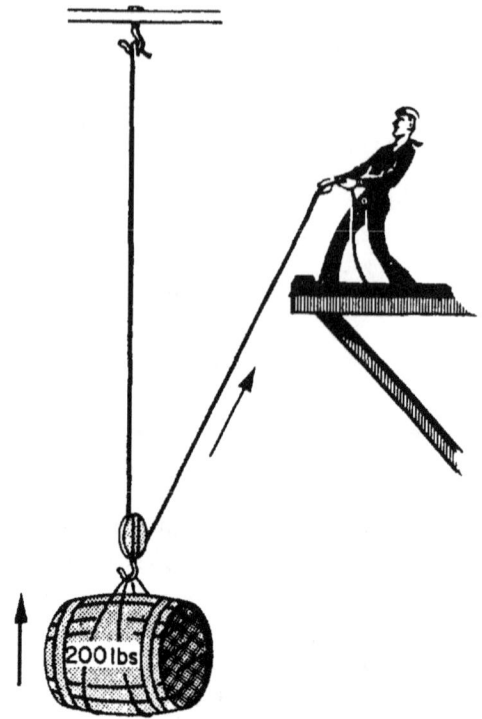

Figure 2-5.-It's 2 to 1.

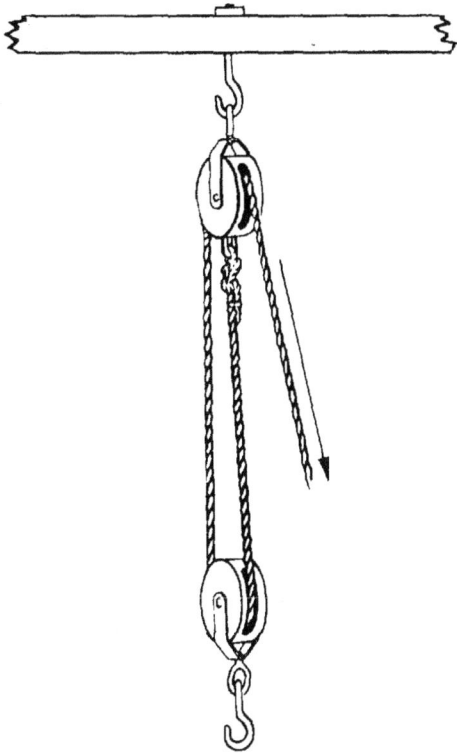

Figure 2-6.-A gun tackle.

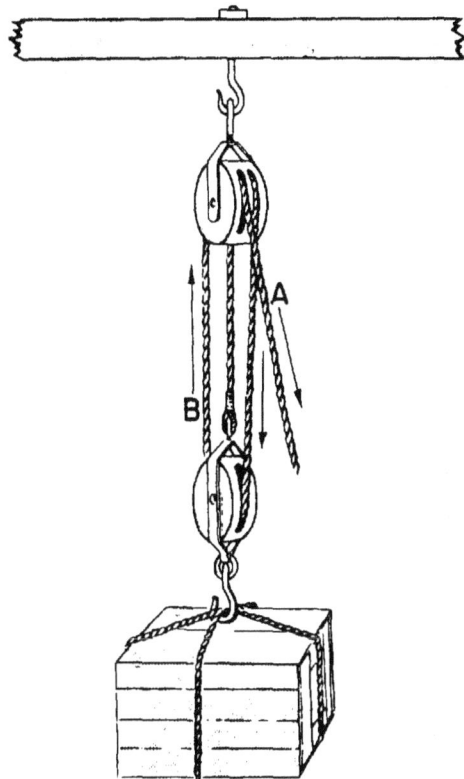

Figure 2-7.-A luff tackle.

The single movable block in this setup is a second-class lever. See figure 2-5. Your effort (E) acts upward upon the arm (EF), which is the diameter of the sheave. The resistance (R) acts downward on the arm (FR), which is the radius of the sheave. Since the diameter is twice the radius, the mechanical advantage is 2.

When the effort at E moves up 2 feet, the load at R is raised only 1 foot. That's something to remember about blocks and falls—if you are actually getting a mechanical advantage from the system. The length of rope that passes through your hands is greater than the distance that the load is raised. However, if you can lift a big load with a small effort, you don't care how much rope you have to pull.

The sailor in figure 2-4 is in an awkward position to pull. If he had another single block handy, he could use it to change the direction of the pull, as in figure 2-6. This second arrangement is known as a gun tackle. Because the second block is fixed, it merely changes the direction of pull—and the mechanical advantage of the whole system remains 2.

You can arrange blocks in several ways, depending on the job to be done and the mechanical advantage you want to get. For example, a luff tack consists of a double block and a single block, rigged as in figure 2-7. Notice that the weight is suspended by the three parts of rope that extend from the movable single block. Each part of the rope carries its share of the load. If the crate weighs 600 pounds, then each of the three parts of the rope supports its share—200 pounds. If there's a pull of 200 pounds downward on rope B, you will have to pull downward with a force of 200 pounds on A to counterbalance the pull on B. Neglecting the friction in the block, a pull of 200 pounds is all that is necessary to raise the crate. The mechanical advantage is:

$$\text{M.A.} = \frac{R}{E} = \frac{600}{200} = 3$$

Here's a good tip. If you count the number of parts of rope going to and from the movable block you can figure the mechanical advantage at a glance. This simple rule will help you to approximate the mechanical advantage of most tackles you see in the Navy.

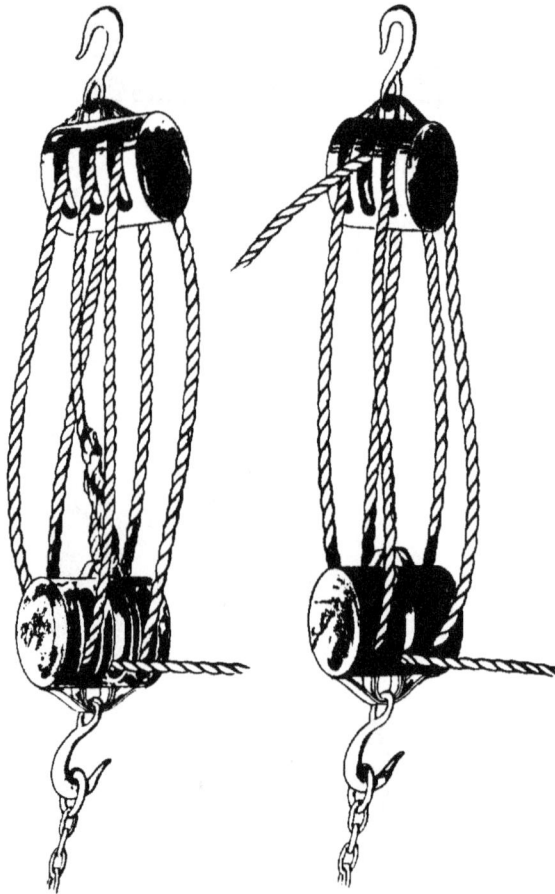

Figure 2-8.-Some other tackles.

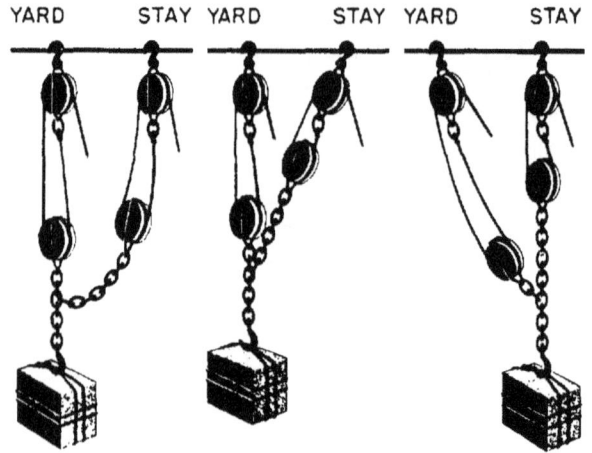

Figure 2-9.-A yard and stay tackle.

Many combinations of single-, double-, and triple-sheave blocks are possible. Two of these combinations are shown in figure 2-8.

You can secure the dead end of the fall to the movable block. The advantage is increased by 1. Notice that this is done in figure 2-7. That is a good point to remember. Remember, also, that the strength of your fall—rope—is a limiting factor in any tackle. Be sure your fall will carry the load. There is no point in rigging a 6-fold purchase that carries a 5-ton load with two triple blocks on a 3-inch manila rope attached to a winch. The winch could take it, but the rope couldn't.

Now for a review of the points you have learned about blocks, and then to some practical applications aboard ship:

With a single fixed block the only advantage is the change of direction of the pull. The mechanical advantage is still 1.

A single movable block gives a mechanical advantage of 2.

Many combinations of single, double, and triple blocks can be rigged to give greater advantages.

Remember that the number of parts of the fall going to and from the movable block tells you the approximate mechanical advantage of the tackle.

If you fix the dead end of the fall to the movable block you increase the mechanical advantage by one 1.

APPLICATIONS AFLOAT AND ASHORE

We use blocks and tackle for lifting and moving jobs afloat and ashore. The five or six basic combinations are used over and over in many situations. Cargo is loaded aboard, and depth charges are stored in their racks. You lower lifeboats over the side with this machine. We can swing heavy machinery, guns, and gun mounts into position with blocks and tackle. In a thousand situations, sailors find this machine useful and efficient.

We use yard and stay tackles aboard ship to pick up a load from the hold and swing it onto the deck. We use yard and stay tackles to shift any load a short distance. Figure 2-9 shows you how to pick a load by the yard tackle. The stay tackle is left slack. After raising the load to the height necessary to clear obstructions, you take up on the stay tackle and ease off on the yard fall. A glance at the rig tells you that the mechanical advantage of each of these tackles is only 2. You may think it's hard work to rig a yard and stay tackle when the small advantage is to move a 400-pound crate along the deck. However, a few minutes spent in rigging may save many unpleasant hours with a sprained back.

If you want a high mechanical advantage, a luff upon luff is a good rig for you. You can raise heavy loads with this setup. Figure 2-10 shows you what a luff upon

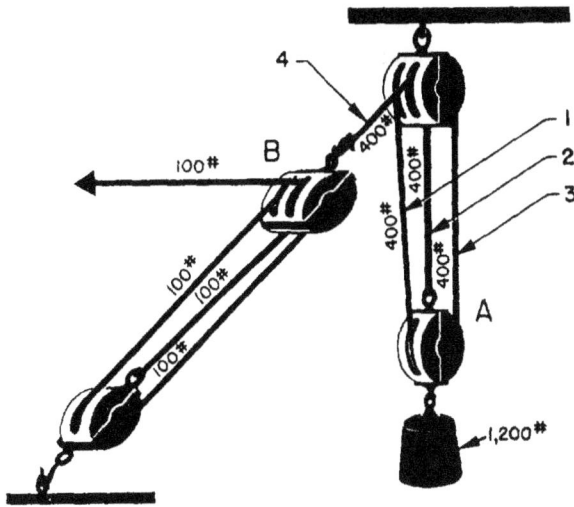

Figure 2-10.-Luff upon luff.

luff rig looks like. If you apply the rule by which you count the parts of the fall going to and from the movable blocks, you find that block A gives a mechanical advantage of 3 to 1. Block B has four parts of fall running to and from it, a mechanical advantage of 4 to 1. The mechanical advantage of those obtained from A is multiplied four times in B. The overall mechanical advantage of a luff upon luff is the product of the two mechanical advantages—or 12.

Don't make the mistake of adding mechanical advantages. Always multiply them.

You can easily figure out the mechanical advantage for the apparatus shown in figure 2-10. Suppose the load weighs 1,200 pounds. The support is by parts 1, 2, and 3 of the fall running to and from block A. Each part must be supporting one-third of the load, or 400 pounds. If part 3 has a pull of 400 pounds on it, part 4—made fast to block B—also has a 400-pound pull on it. There are four parts of the second fall going to and from block B. Each of these takes an equal part of the 400—pound pull. Therefore, the hauling part requires a pull of only 1/4 x 400, or 100 pounds. So, here you have a 100-pound pull raising a 1,200-pound load. That's a mechanical advantage of 12.

In shops ashore and aboard ship, you are almost certain to run into a chain hoist, or differential pulley. Ordinarily, you suspend these hoists from overhead trolleys. You use them to lift heavy objects and move them from one part of the shop to another.

To help you to understand the operation of a chain hoist, look at the one in figure 2-11. Assume that you grasp the chain (E) and pull until the large wheel (A) has

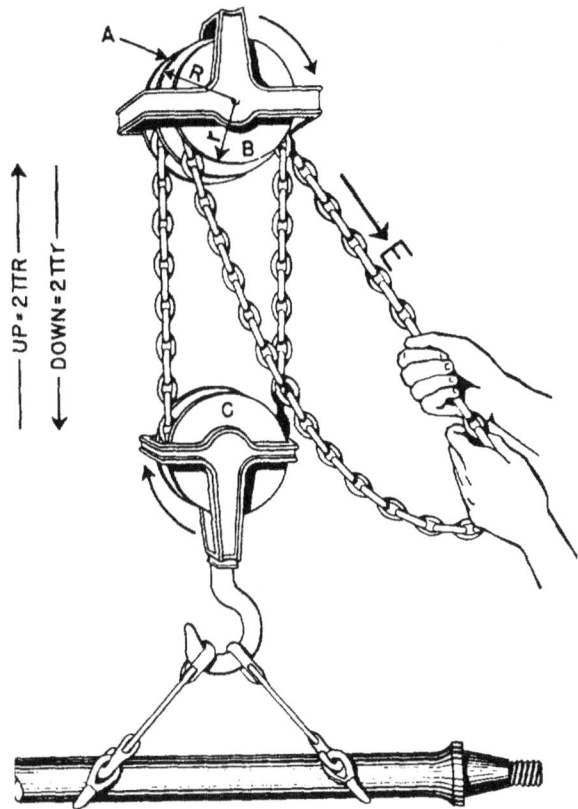

Figure 2-11.—A chain hoist.

turned around once. Then the distance through which your effort has moved is equal to the circumference of that wheel, or $2\pi r$. Again, since C is a single movable block the downward movement of its center will be equal to only one-half the length of the chain fed to it, or πr.

Of course, C does not move up a distance πR and then move down a distance πr. Actually, its steady movement upward is equal to the difference between the two, or ($\pi R - \pi r$). Don't worry about the size of the movable pulley (C). It doesn't enter into these calculations. Usually, its diameter is between that of A and that of B.

The mechanical advantage equals the distance that moves the effort (E). It's divided by the distance that moves the load. We call this the velocity ratio, or theoretical mechanical advantage (T.M.A.). It is theoretical because the frictional resistance to the movement of mechanical parts is left out. In practical uses, all moving parts have frictional resistance.

The equation for theoretical mechanical advantage may be written

$$\frac{\text{Distance effort moves}}{\text{Distance resistance moves}} ;$$

and in this case,

$$\text{T.M.A.} = \frac{2\pi R}{\pi R - \pi r} \cdot \frac{2R}{(R - r)}.$$

If A is a large wheel and B is a little smaller, the value of $2R$ becomes large and then $(R - r)$ becomes small. Then you have a large number for

$$\frac{2R}{(R - r)}$$

which is the theoretical mechanical advantage.

You can lift heavy loads with chain hoists. To give you an idea of the mechanical advantage of a chain hoist, suppose the large wheel has a radius (R) of 6 inches and the smaller wheel a radius (r) of 5 3/4 inches. What theoretical mechanical advantage would you get? Use the formula

$$\text{T.M.A.} = \frac{2R}{R - r}.$$

Then substitute the numbers in their proper places, and solve

$$\text{T.M.A.} = \frac{2 \times 6}{6 - 5\frac{3}{4}} = \frac{12}{\frac{1}{4}} = 48$$

Since the friction in this type of machine is considerable, the actual mechanical advantage is not as high as the theoretical mechanical advantage. For example, that theoretical mechanical advantage of 48 tells you that with a 1-pound pull you can lift a 48-pound load. However, actually your 1-pound pull might only lift a 20-pound load. You will use the rest of your effort in overcoming the friction.

SUMMARY

The most important point to remember about block and tackle is that they are simple machines. And simple machines multiply effort or change its direction. You should also remember the following points:

A pulley is a grooved wheel that turns by the action of a rope in the groove.

There are different types of pulleys. Pulleys are either fixed or movable.

You attach a fixed pulley to one place. The fixed pulley helps make work easier by changing the direction of the effort.

You hook a movable pulley to the object you are lifting. As you pull, the object and the pulley move together. This pulley does not change the direction of the effort, but it does multiply the effort.

You can use fixed and movable pulleys together to get a large mechanical advantage (M.A.).

CHAPTER 3

THE WHEEL AND AXLE

CHAPTER LEARNING OBJECTIVES

Upon completion of this chapter, you should be able to do the following:

- *Explain the advantage of the wheel and axle.*

Have you ever tried to open a door when the knob was missing? If you have, you know that trying to twist that small four-sided shaft with your fingers is tough work. That gives you some appreciation of the advantage you get by using a knob. The doorknob is an example of a simple machine called a wheel and axle.

The steering wheel on an automobile, the handle of an ice cream freezer, and a brace and bit are all examples of a simple machine. All of these devices use the wheel and axle to multiply the force you exert. If you try to turn a screw with a screwdriver and it doesn't turn, stick a screwdriver bit in the chuck of a brace. The screw will probably go in with little difficulty.

There's something you'll want to get straight right at the beginning. The wheel-and-axle machine consists of a wheel or crank rigidly attached to the axle, which turns with the wheel. Thus, the front wheel of an automobile is not a wheel-and-axle machine because the axle does not turn with the wheel.

MECHANICAL ADVANTAGE

How does the wheel-and-axle arrangement help to magnify the force you exert? Suppose you use a screwdriver bit in a brace to drive a stubborn screw. Look at figure 3-1, view A. You apply effort on the handle that moves in a circular path, the radius of which is 5 inches. If you apply a 10-pound force on the handle, how much force will you exert against the resistance at the screw? Assume the radius of the screwdriver blade is 1/4 inch. You are really using the brace as a second-class lever—see figure 3-1, view B. You can find the size of the resistance by using the formula

$$\frac{L}{1} = \frac{R}{E}.$$

In that

L = radius of the circle through which the handle turns,

1 = one-half the width of the edge of the screwdriver blade,

R = force of the resistance offered by the screw,

E = force of effort applied on the handle.

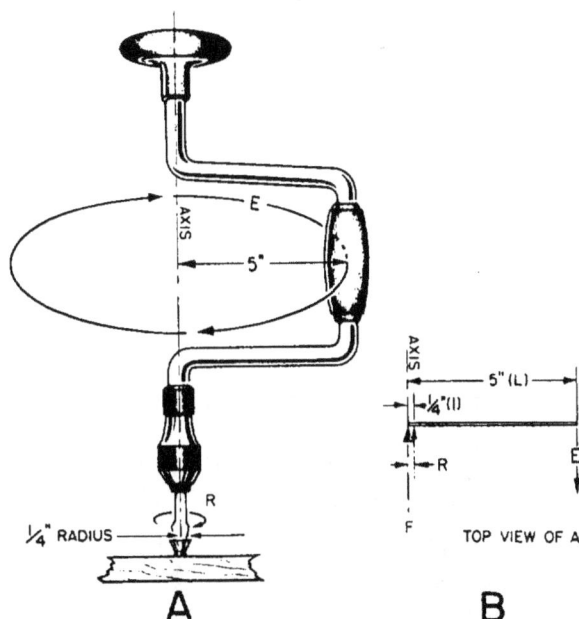

Figure 3-1.-It magnifies your effort.

3-1

Substituting in the formula and solving:

$$\frac{5}{\frac14} = \frac{R}{10}$$

$$R = \frac{5 \times 10}{\frac14}$$

$$= 5 \times 10 \times 4$$

$$= 200 \text{ lb.}$$

This means that the screwdriver blade will turn the screw with a force of 200 pounds. The relationship between the radius of the diameters or the circumferences of the wheel and axle tells you how much mechanical advantage you can get.

Take another situation. You raise the old oaken bucket, figure 3-2, using a wheel-and-axle arrangement. If the distance from the center of the axle to the handle is 8 inches and the radius of the drum around which the rope is wound is 2 inches, then you have a theoretical mechanical advantage of 4. That's why these rigs were used.

MOMENT OF FORCE

In several situations you can use the wheel-and-axle to speed up motion. The rear-wheel sprocket of a bike, along with the rear wheel itself, is an example. When you are pedaling, the sprocket is attached to the wheel; so the combination is a true wheel-and-axle machine. Assume that the sprocket has a circumference of 8 inches, and the wheel circumference is 80 inches. If you turn the sprocket at a rate of one revolution per second, each sprocket tooth moves at a speed of 8 inches per second. Since the wheel makes one revolution for each revolution made by the sprocket, any point on the tire must move through a distance of 80 inches in 1 second. So, for every 8-inch movement of a point on the sprocket, you have moved a corresponding point on the wheel through 80 inches.

Since a complete revolution of the sprocket and wheel requires only 1 second, the speed of a point on the circumference of the wheel is 80 inches per second, or 10 times the speed of a tooth on the sprocket.

(NOTE: Both sprocket and wheel make the same number of revolutions per second, so the speed of turning for the two is the same.)

Here is an idea that you will find useful in understanding the wheel and axle, as well as other machines. You probably have noticed that the force you apply to a lever starts to turn or rotate it about the fulcrum. You also know that a sheave on a fall starts to rotate the sheave of the block. Also when you turn the steering wheel of a car, it starts to rotate the steering column. Whenever you use a lever, or a wheel and axle, your effort on the lever arm or the rim of the wheel causes it to rotate about the fulcrum or the axle in one direction or another. If the rotation occurs in the same direction as the hands of a clock, we call that direction clockwise. If the rotation occurs in the opposite direction from that of the hands of a clock, we call that direction of rotation counterclockwise. A glance at figure 3-3 will make clear the meaning of these terms.

The force acting on the handle of a carpenter's brace depends not only on the amount of that force, but also on the distance from the handle to the center of rotation. This is known as a moment of force, or a torque (pronounced tork). Moment of force and torque have the same meaning.

Look at the effect of the counterclockwise movement of the capstan bar in figure 3-4. Here the amount of the effort is designated E_1 and the distance from the point where you apply the force to the center

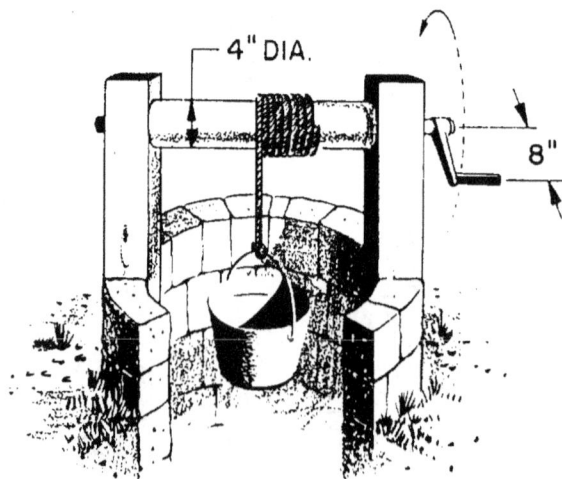

Figure 3-2.-The old oaken bucket.

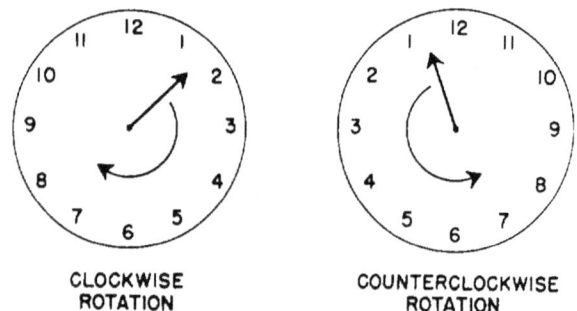

Figure 3-3.-Directions of rotation.

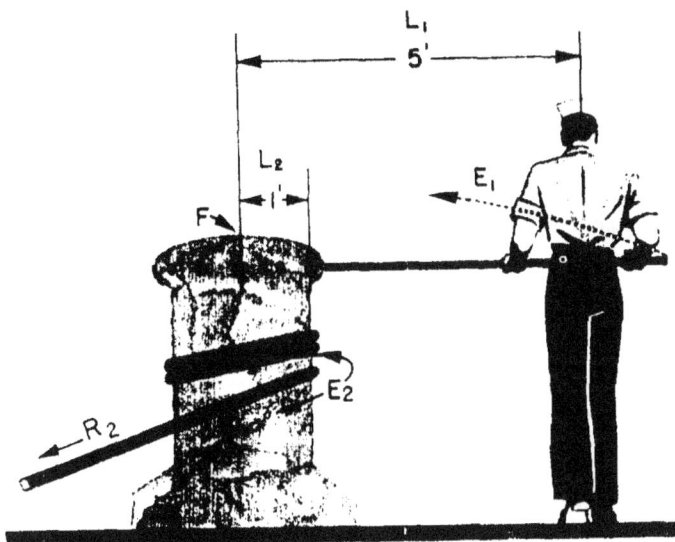

Figure 3-4.-Using the capstan.

of the axle is L_1. Then, $E_1 x L_1$ is the moment of force. You'll notice that this term includes both the amount of the effort and the distance from the point of application of effort to the center of the axle. Ordinarily, you measure the distance in feet and the applied force in pounds.

Therefore, you measure moments of force in foot-pounds (ft-lb). A moment of force is frequently called a moment.

By using a longer capstan bar, the sailor in figure 3-4 can increase the effectiveness of his push without making a bigger effort. If he applied his effort closer to the head of the capstan and used the same force, the moment of force would be less.

BALANCING MOMENTS

You know that the sailor in figure 3-4 would land flat on his face if the anchor hawser snapped. As long as nothing breaks, he must continue to push on the capstan bar. He is working against a clockwise moment of force that is equal in magnitude, but opposite in direction, to his counterclockwise moment of force. The resisting moment, like the effort moment, depends on two factors. In the case of resisting moment, these factors are the force (R_2) with which the anchor pulls on the hawser and the distance (L_2) from the center of the capstan to its rim. The existence of this resisting force would be clear if the sailor let go of the capstan bar. The weight of the anchor pulling on the capstan would cause the whole works to spin rapidly in a clockwise direction—and good-bye

anchor! The principle involved here is that whenever the counterclockwise and the clockwise moments of force are in balance, the machine either moves at a steady speed or remains at rest.

This idea of the balance of moments of force can be summed up by the expression

$$\text{CLOCKWISE MOMENTS} = \text{COUNTERCLOCKWISE MOMENTS}$$

Since a moment of force is the product of the amount of the force times the distance the force acts from the center of rotation, this expression of equality may be written

$$E_1 \text{ x } L_1 = E_2 \times L_2,$$

in that

E_1 = force of effort,

L_1 = distance from fulcrum or axle to point where you apply force,

E_2 = force of resistance, and

L_2 = distance from fulcrum or center axle to the point where you apply resistance.

EXAMPLE 1

Put this formula to work on a capstan problem. You grip a single capstan bar 5 feet from the center of a capstan head with a radius of 1 foot. You have to lift a 1/2-ton anchor. How big of a push does the sailor have to exert?

First, write down the formula

$$E_1 \times L_1 = E_2 \times L_2.$$

Here

L_1 = 5

E_2 = 1,000 pounds, and

L_2 = 1.

Substitute these values in the formula, and it becomes:

$$E_1 \times 5 = 1,000 \text{ x } 1$$

and

$$E_1 = \frac{1,000}{5} = 200 \text{ pounds}$$

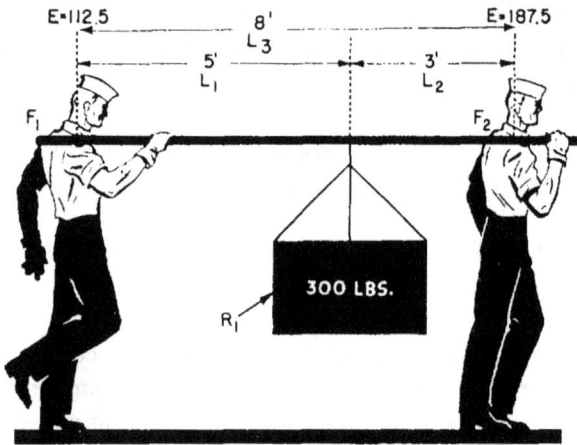

Figure 3-5.-A practical application.

Example 2

Consider now the sad case of Slim and Sam, as illustrated in figure 3-5. Slim has suggested that they carry the 300-pound crate slung on a handy 10-foot pole. He was smart enough to slide the load up 3 feet from Sam's shoulder.

Here's how they made out. Use Slim's shoulder as a fulcrum (F_1). Look at the clockwise movement caused by the 300-pound load. That load is 5 feet away from Slim's shoulder. If R_1 is the load, and L_1 the distance from Slim's shoulder to the load, the clockwise moment (M_A) is

$$M_A = R_1 \times L_1 = 300 \times 5 = 1,500 \text{ ft-lb.}$$

With Slim's shoulder still acting as the fulcrum, the resistance of Sam's effort causes a counterclockwise moment (M_B) acting against the load moment. This counterclockwise moment is equal to Sam's effort (E_2) times the distance (L_3) from his shoulder to the fulcrum (F_1) at Slim's shoulder. Since $L_2 = 8$ ft, the formula is

$$M_B = E_2 \text{ x } L_3 = E_2 \times 8 = 8E_2$$

There is no rotation, so the clockwise moment and the counterclockwise moment are equal. $MA = M_B$. Hence

$$1,500 = 8E_2$$

$$E_2 = \frac{1,500}{8} = 187.5 \text{ pounds.}$$

So poor Sam is carrying 187.5 pounds of the 330-pound load.

What is Slim carrying? The difference between 300 and 187.5 = 112.5 pounds, of course! You can check your answer by the following procedure.

This time, use Sam's shoulder as the fulcrum (F_2). The counterclockwise moment (M_c) is equal to the 300-pound load (R_1) times the distance $(L_2 = 3$ feet) from Sam's shoulder. M_c 300 x 3 = 900 foot-pounds. The clockwise moment (m_D) is the result of Slim's lift (E_1) acting at a distance (L_3) from the fulcrum. $L_3 = 8$ feet. Again, since counterclockwise moment equals clockwise moment, you have

$$900 = E_1 \times 8$$

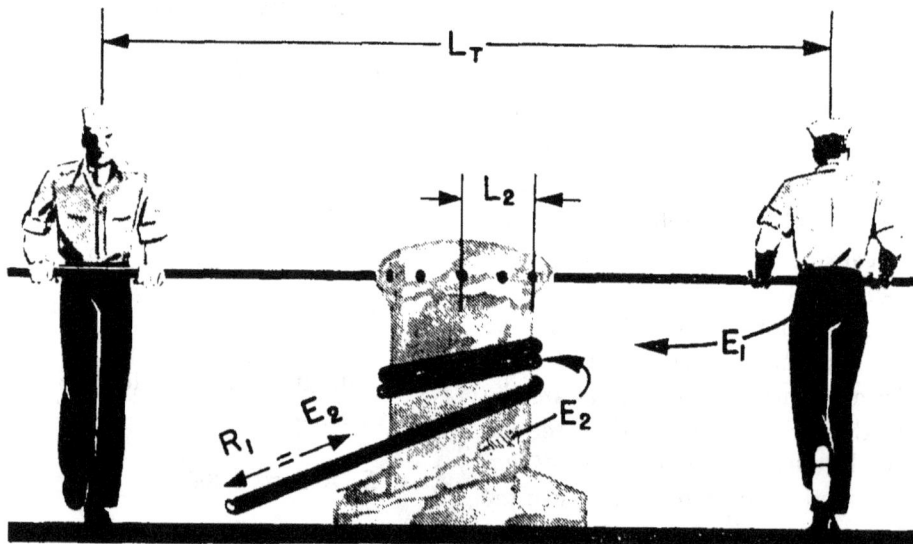

Figure 3-6.-A couple.

Figure 3-7.-Valves.

and

$$E_1 = \frac{900}{8} = 112/5 \text{ pounds}$$

Slim, the smart sailor, has to lift only 112.5 pounds. There's a sailor who really puts his knowledge to work.

THE COUPLE

Take a look at figure 3-6. It's another capstan-turning situation. To increase an effective effort, place a second capstan bar opposite the first and another sailor can apply a force on the second bar. The two sailors in figure 3-6 will apparently be pushing in opposite directions. Since they are on opposite sides of the axle, they are actually causing rotation in the same direction. If the two sailors are pushing with equal force, the moment of force is twice as great as if only one sailor were pushing. This arrangement is known technically as a couple.

You will see that the couple is a special example of the wheel and axle. The moment of force is equal to the product of the total distance (L_n between the two points and the force (E_1) applied by one sailor. The equation for the couple may be written

$$E_1 \times L_T = E_2 \times L_2$$

APPLICATIONS AFLOAT AND ASHORE

A trip to the engine room important the wheel and axle makes you realize how is on the modern ship.

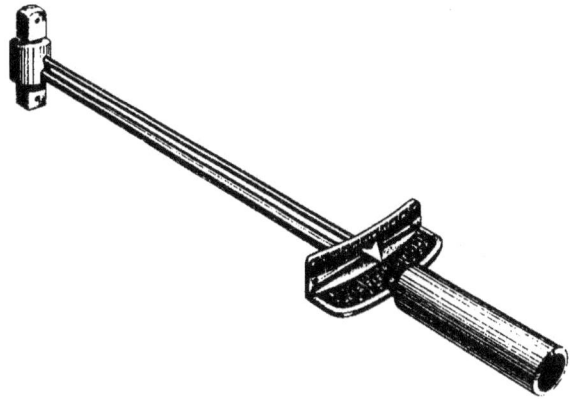

Figure 3-8.—A simple torque wrench.

Everywhere you look you see wheels of all sizes and shapes. We use most of them to open and close valves quickly. One common type of valve is shown in figure 3-7. Turning the wheel causes the threaded stem to rise and open the valve. Since the valve must close watertight, airtight, or steamtight, all the parts must fit snugly. To move the stem on most valves without the aid of the wheel would be impossible. The wheel gives you the necessary mechanical advantage.

You've handled enough wrenches to know that the longer the handle, the tighter you can turn a nut. Actually, a wrench is a wheel-and-axle machine. You can consider the handle as one spoke of a wheel and the place where you take hold of the handle as a point on the rim. You can compare the nut that holds in the jaws of the wrench to the axle.

You know that you can turn a nut too tight and strip the threads or cause internal parts to seize. This is especially true when you are taking up on bearings. To make the proper adjustment, you use a torque wrench. There are several types. Figure 3-8 shows you one that is very simple. When you pull on the handle, its shaft bends. The rod fixed on the pointer does not bend. The pointer shows on the scale the torque, or moment of force, that you are exerting. The scale indicates pounds, although it is really measuring foot-pounds to torque. If the nut is to be tightened by a moment of 90 ft-1b, you pull until the pointer is opposite the number 90 on the scale. The servicing or repair manual on an engine or piece of machinery tells you what the torque—or moment of force—should be on each set of nuts or bolt.

The gun pointer uses a couple to elevate and depress the gun barrel. He cranks away at a handwheel that has two handles. The right-hand handle is on the opposite side of the axle from the left-hand handle—180° apart.

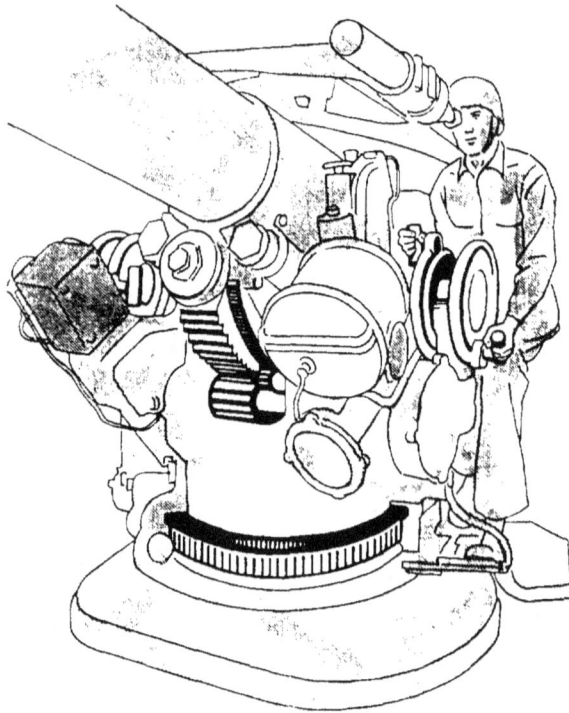

Figure 3-9.-A pointer's handwheel.

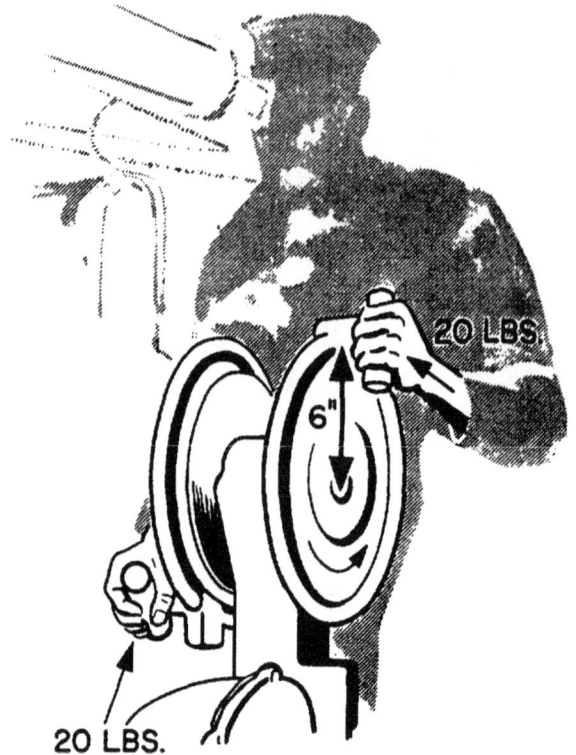

Figure 3-10.-Developing a torque.

Look at figure 3-9. When this gun pointer pulls on one handle and pushes on the other, he's producing a couple. If he cranks only with his right hand, he no longer has a couple—just a simple first-class lever! And he'd have to push twice as hard with one hand.

A system of gears-a gear train-transmits the motion to the barrel. A look at figure 3-10 will help you to figure the forces involved. The radius of the wheel is 6 inches—1/2 foot-and turns each handle with a force of 20 pounds. The moment on the top that rotates the wheel in a clockwise direction is equal to 20 x 1/2 = 10 ft-lb. The bottom handle also rotates the wheel in the same direction with an equal moment. Thus, the total twist or torque on the wheel is 10 + 10 = 20 ft-lb. To get the same moment with one hand, apply a 20-pound force. The radius of the wheel would have to be twice as much—12 inches—or one foot. The couple is a convenient arrangement of the wheel-and-axle machine.

SUMMARY

Here is a quick review of the wheel and axle-facts you should have straight in your mind:

A wheel-and-axle machine has the wheel fixed rigidly to the axle. The wheel and the axle turn together.

Use the wheel and axle to magnify your effort or to speed it up.

You call the effect of a force rotating an object around an axis or fulcrum a moment of force, or simply a moment.

When an object is at rest or is moving steadily, the clockwise moments are just equal and opposite to the counterclockwise moments.

Moments of force depend upon two factors: (1) the amount of the force and (2) the distance from the fulcrum or axis to the point where the force is applied.

When you apply two equal forces at equal distances on opposite sides of a fulcrum and move those forces in opposite directions so they both tend to cause rotation about the fulcrum, you have a couple.

CHAPTER 4

THE INCLINED PLANE AND THE WEDGE

CHAPTER LEARNING OBJECTIVES

Upon completion of this chapter, you should be able to do the following:

- *Summarize the advantage of the barrel roll and the wedge.*

You have probably watched a driver load barrels on a truck. He backs the truck up to the curb. The driver then places a long double plank or ramp from the sidewalk to the tailgate, and then rolls the barrel up the ramp. A 32-gallon barrel may weigh close to 300 pounds when full, and it would be a job to lift one up into the truck. Actually, the driver is using a simple machine called the inclined plane. You have seen the inclined plane used in many situations. Cattle ramps, a mountain highway and the gangplank are familiar examples.

The inclined plane permits you to overcome a large resistance, by applying a small force through a longer distance when raising the load. Look at figure 4-1. Here you see the driver easing the 300-pound barrel up to the bed of the truck, 3 feet above the sidewalk. He is using a plank 9 feet long. If he didn't use the ramp at all, he'd have to apply 300-pound force straight up through the 3-foot distance. With the ramp, he can apply his effort over the entire 9 feet of the plank as he rolls the barrel to a height of 3 feet. It looks as if he could use a force only three-ninths of 300, or 100 pounds, to do the job. And that is actually the situation.

Here's the formula. Remember it from chapter 1?

$$\frac{L}{1} = \frac{R}{E}$$

In which

L = length of the ramp, measured along the slope,

1 = height of the ramp,

R = weight of the object to be raised, or lowered,

E = force required to raise or lower the object.

Now apply the formula this problem:

In this case,

L = 9ft,

1 = 3 ft, and

R = 300 lb.

By substituting these values in the formula, you get

$$\frac{9}{3} = \frac{300}{E}$$

$9E$ = 900

E = 100 pounds.

Since the ramp is three times as long as its height, the mechanical advantage is three. You find the theoretical mechanical advantage by dividing the total distance of the effort you exert by the vertical distance the load is raised or lowered.

THE WEDGE

The wedge is a special application of the inclined plane. You have probably used wedges. Abe Lincoln used a wedge to help him split logs into rails for fences. The blades of knives, axes, hatchets, and chisels act as wedges when they are forced into apiece of wood. The wedge is two inclined planes set base-to-base. By

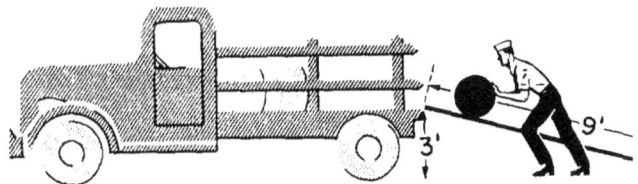

Figure 4-1.—An inclined plane.

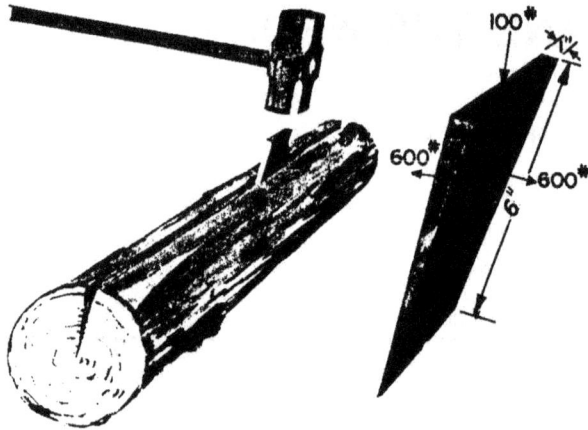

Figure 4-2.-A wedge.

Figure 4-3.—The gangplank is an inclined plane.

driving the wedge full-length into the material to cut or split, you force the material apart a distance equal to the width of the broad end of the wedge. See figure 4-2.

Long, slim wedges give high mechanical advantage. For example, the wedge of figure 4-2 has a mechanical advantage of six. The greatest value of the wedge is that you can use it in situations in which other simple machines won't work. Imagine the trouble you'd have trying to pull a log apart with a system of pulleys.

APPLICATIONS AFLOAT AND ASHORE

A common use of the inclined plane in the Navy is the gangplank. Going aboard the ship by gangplank illustrated in figure 4-3, is easier than climbing a sea ladder. You appreciate the mechanical advantage of the gangplank even more when you have to carry your seabag or a case of sodas aboard.

Remember that hatch dog in figure 1-10? The use of the dog to secure a door takes advantage of the lever principle. If you look sharply, you can see that the dog seats itself on a steel wedge welded to the door. As the dog slides upward along this wedge, it forces the door tightly shut. This is an inclined plane, with its length about eight times its thickness. That means you get a theoretical mechanical advantage of eight. In chapter 1, you figured that you got a mechanical advantage of four from the lever action of the dog. The overall mechanical advantage is 8 x 4, or 32, neglecting friction. Not bad for such a simple gadget, is it? Push down with 50 pounds heave on the handle and you squeeze the door

shut with a force of 1,600 pounds on that dog. You'll find the damage-control parties using wedges by the dozen to shore up bulkheads and decks. A few sledgehammer blows on a wedge will quickly and firmly tighten up the shoring.

Chipping scale or paint off steel is a tough job. How-ever, you can make the job easier with a compressed-air chisel. The wedge-shaped cutting edge of the chisel gets in under the scale or the paint and exerts a large amount of pressure to lift the scale or paint layer. The chisel bit is another application of the inclined plane.

SUMMARY

This chapter covered the following points about the inclined plane and the wedge:

The inclined plane is a simple machine that lets you raise or lower heavy objects by applying a small force over a long distance.

You find the theoretical mechanical advantage of the inclined plane by dividing the length of the ramp by the perpendicular height of the load that is raised or lowered. The actual mechanical advantage is equal to the weight of the resistance or load, divided by the force that must be used to move the load up the ramp.

The wedge is two inclined planes set base-to-base. It finds its greatest use in cutting or splitting materials.

CHAPTER 5

THE SCREW

CHAPTER LEARNING OBJECTIVES

Upon completion of this chapter, you should be able to do the following:

- *State the uses of the screw.*

- *Explain the use of the jack.*

- *Discuss the use of the micrometer*

The screw is a simple machine that has many uses. The vise on a workbench makes use of the mechanical advantage (M.A.) of the screw. You get the same advantage using glued screw clamps to hold pieces of furniture together, a jack to lift an automobile, or a food processor to grind meat.

A screw is a modification of the inclined plane. Cut a sheet of paper in the shape of a right triangle and you have an inclined plane. Wind this paper around a pencil, as in figure 5-1, and you can see that the screw is actually an inclined plane wrapped around a cylinder. As you turn the pencil, the paper is wound up so that its hypotenuse forms a spiral thread. The pitch of the screw and paper is the distance between identical points on the same threads measured along the length of the screw.

THE JACK

To understand how the screw works, look at figure 5-2. Here you see the type of jack screw used to raise a house or apiece of heavy machinery. Notice that the jack has a lever handle; the length of the handle is equal to r.

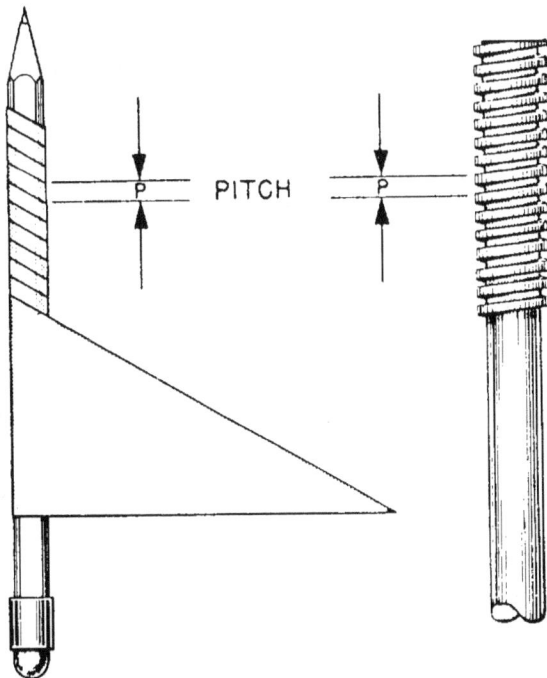

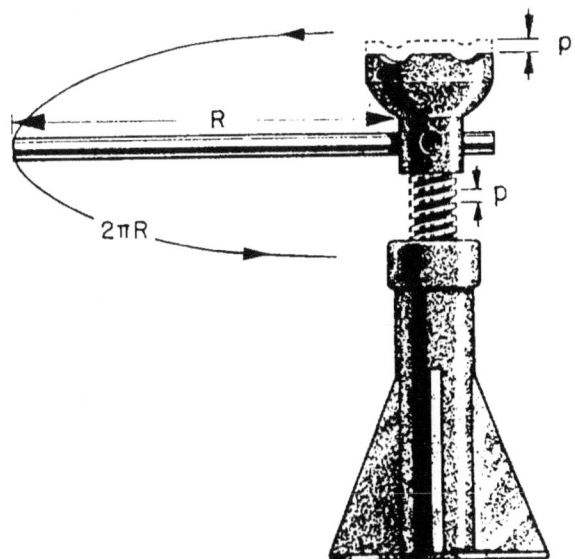

Figure 5-1.—A screw is an inclined plane in spiral form.

Figure 5-2.-A jack screw.

If you pull the lever handle around one turn, its outer end has described a circle. The circumference of this circle is equal to 2π. (Remember that π equals 3.14, or $^{22}/_7$.) That is the distance you must apply the effort of the lever arm.

At the same time, the screw has made one revolution, raising its height to equal its pitch (*p*). You might say that one full thread has come up out of the base. At any rate, the load has risen a distance *p*.

Remember that the theoretical mechanical advantage (T.M.A.) is equal to the distance through which you apply the effort or pull, divided by the distance and resistance the load is moved. Assuming a 2-foot, or 24-inch, length for the lever arm and a 1/4-inch pitch for the thread, you can find the theoretical mechanical advantage by the formula

$$\text{T.M.A.} = \frac{2\pi r}{P}$$

in that

    r = length of handle = 24 inches

    p = pitch, or distance between corresponding
        points on successive threads = 1/4 inch.

Substituting,

$$\text{T.M.A.} = \frac{2 \times 3.14 \times 24}{\frac{1}{4}} = \frac{150.72}{\frac{1}{4}} = 602.88.$$

A 50-pound pull on the handle would result in a theoretical lift of 50 x 602 or about 30,000 pounds—15 tons for 50 pounds.

However, jacks have considerable friction loss. The threads are cut so that the force used to overcome friction is greater than the force used to do useful work. If the threads were not cut this way and no friction were present, the weight of the load would cause the jack to spin right back down to the bottom as soon as you released the handle.

THE MICROMETER

In using the jack you exerted your effort through a distance of $2\pi r$, or 150 inches, to raise the screw 1/4 inch. It takes a lot of circular motion to get a small amount of straight line motion from the head of the jack. You will use this point to your advantage in the micrometer; it's a useful device for making accurate small measurements—measurements of a few thousandths of an inch.

In figure 5-3, you see a cutaway view of a micrometer. The thimble turns freely on the sleeve,

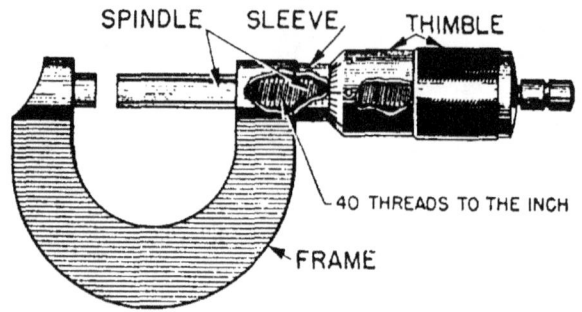

Figure 5-3.-A micrometer.

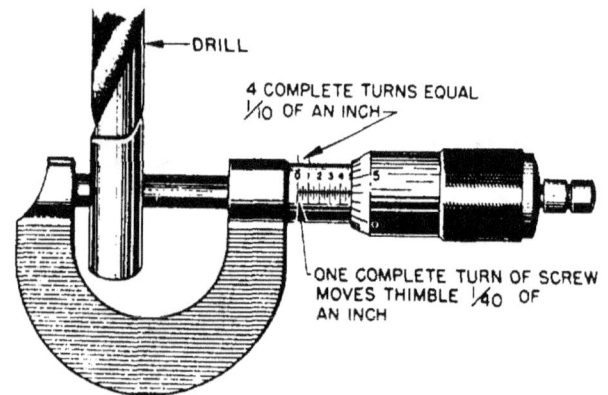

Figure 5-4.—Taking turns.

rigidly attached to the micrometer frame. The spindle attaches to the thimble and is fitted with screw threads that move the spindle and thimble to the right or left in the sleeve when you rotate the thimble. These screw threads are cut 40 threads to the inch. Hence, one turn of the thimble moves the spindle and thimble 1/40 of inch. This represents one of the smallest divisions on the micrometer. Four of these small divisions make 4/40 of an inch, or 1/10 inch. Thus, the distance from 0 to 1 or 1 to 2 on the sleeve represents 1/10, or 0.1, inch.

To allow even finer measurements, the thimble is divided into 25 equal parts. It is laid out by graduation marks around its rim, as shown in figure 5-4. If you turn the thimble through 25 of these equal parts, you have made one complete revolution of the screw. This represents a lengthwise movement of 1/40 of an inch. If you turn the thimble one of these units on its scale, you have moved the spindle a distance of 1/25 of 1/40 inch, or 1/1000 of an inch—0.001 inch.

The micrometer in figure 5-4 reads 0.503 inch, that is the true diameter of the half-inch drill-bit shank measured. This tells you that the diameter of this bit is 0.003 inch greater than its nominal diameter of 1/2 inch—0.5000 inch.

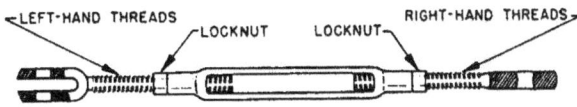

Figure 5-5.—A turnbuckle.

Figure 5-6.-A rigger's vice.

Because you can make accurate measurements with this instrument, it is vital in every machine shop.

APPLICATIONS AFLOAT AND ASHORE

It's a tough job to pull a rope or cable tight enough to get all the slack out of it. However, you can do it by using a turnbuckle. The turnbuckle (fig, 5-5) is an application of the screw. If you turn it in one direction, it takes up the slack in a cable. Turning it the other way allows slack in the cable. Notice that one bolt of the turnbuckle has left-hand threads and the other bolt has right-hand threads. Thus, when you turn the turnbuckle to tighten the line, both bolts tighten up. If both bolts were right-hand thread-standard thread-one would tighten while the other one loosened an equal amount. That would result in no change in cable slack. Most turnbuckles have the screw threads cut to provide a large amount of frictional resistance to keep the turnbuckle from unwinding under load. In some cases, the turnbuckle has a locknut on each of the screws to prevent slipping. You'll find turnbuckles used in a hundred different ways afloat and ashore.

Ever wrestled with a length of wire rope? Obstinate and unwieldy, wasn't it? Riggers have dreamed up tools to help subdue wire rope. One of these tools-the rigger's vise-is shown in figure 5-6. This rigger's vise uses the mechanical advantage of the screw to hold the wire rope in place. The crew splices a thimble-a reinforced loop—onto the end of the cable. Rotating the handle causes the jaw on

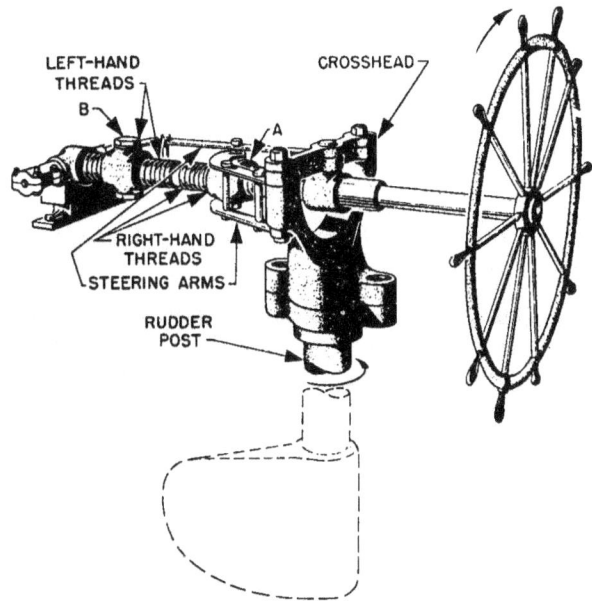

Figure 5-7.—A friction brake.

Figure 5-8.—The screw gives a tremendous mechanical advantage.

that screw to move in or out along its grooves. This machine is a modification of the vise on a workbench. Notice the right-hand and left-hand screws on the left-hand clamp.

Figure 5-7 shows you another use of the screw. Suppose you want to stop a winch with its load suspended in mid-air. To do this, you need a brake. The brakes on most anchor or cargo winches consist of a metal band that encircles the brake drum. The two ends of the band fasten to nuts connected by a screw attached to a handwheel. As you turn the handwheel, the screw pulls the lower end of the band (A) up toward its upper end (B). The huge mechanical advantage of the screw puts the squeeze on the drum, and all rotation of the drum stops.

One type of steering gear used on many small ships and as a spare steering system on some larger ships is the screw gear. Figure 5-8 shows you that the

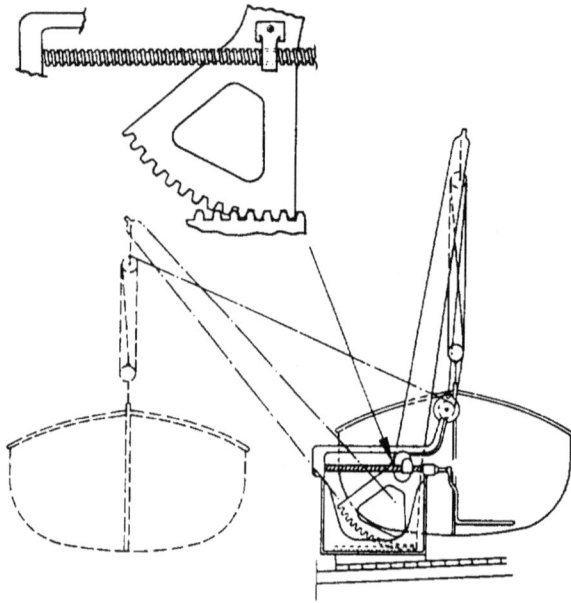

Figure 5-9.—The quadrant davit.

wheel turns a long threaded shaft. Half the threads—those nearer the wheel end of this shaft-are right-hand threads. The other half of the threads-those farther from the wheel—are left-hand threads. Nut A has a right-hand thread, and nut B has a left-hand thread. Notice that two steering arms connect the crosshead to the nuts; the crosshead turns the rudder. If you stand in front of the wheel and turn it in a clockwise direction—to your right—arm A moves forward and arm B moves backward. That turns the rudder counterclockwise, so the ship swings in the direction you turn the wheel. This steering mechanism has a great mechanical advantage.

Figure 5-9 shows you another practical use of the screw. The quadrant davit makes it possible for two men

to put a large lifeboat over the side with little effort. The operating handle attaches to a threaded screw that passes through a traveling nut. Cranking the operating handle in a counterclockwise direction (as you face outboard), the nut travels outward along the screw. The traveling nut fastens to the davit arm by a swivel. The davit arm and the boat swing outboard as a result of the outward movement of the screw. The thread on that screw is the self-locking type; if you let go of the handle, the nut remains locked in position.

SUMMARY

You have learned the following basic information about the screw from this chapter; now notice the different ways the Navy uses this simple machine:

The screw is a modification of the inclined plane—modified to give you a high mechanical advantage.

The theoretical mechanical advantage of the screw can be found by the formula

$$\text{T.M.A.} = \frac{2\pi r}{p}.$$

As in all machines, the actual mechanical advantage equals the resistance divided by the effort.

In many applications of the screw, you make use of the large amount of friction that is commonly present in this simple machine.

By using the screw, you reduce large amounts of circular motion to very small amounts of straight-line motion.

CHAPTER 6

GEARS

CHAPTER LEARNING OBJECTIVES

Upon completion of this chapter, you should be able to do the following:

● *Compare the types of gears and their advantages.*

Did you ever take a clock apart to see what made it tick? Of course you came out with some parts left over when you got it back together again. And they probably included a few gear wheels. We use gears in many machines. Frequently the gears are hidden from view in a protective case filled with grease or oil, and you may not see them.

An eggbeater gives you a simple demonstration of the three jobs that gears do. They can change the direction of motion, increase or decrease the speed of the applied motion, and magnify or reduce the force that you apply. Gears also give you a positive drive. There can be, and usually is, creep or slip in a belt drive. However, gear teeth are always in mesh, so there can be no creep and slip.

Follow the directional changes in figure 6-1. The crank handle turns in the direction shown by the arrow—clockwise—when viewed from the right. The 32 teeth on the large vertical wheel (A) mesh with the 8 teeth on the right-hand horizontal wheel (B), which rotates as shown by the arrow. Notice that as B turns in a clockwise direction, its teeth mesh with those of wheel C and cause wheel C to revolve in the opposite direction. The rotation of the crank handle has been transmitted by gears to the beater blades, which also rotate.

Now figure out how the gears change the speed of motion. There are 32 teeth on gear A and 8 teeth on gear B. However, the gears mesh, so that one complete revolution of A results in four complete revolutions of gear B. And since gears B and C have the same number of teeth, one revolution of B results in one revolution of C. Thus, the blades revolve four times as fast as the crank handle.

In chapter 1 you learned that third-class levers increase speed at the expense of force. The same happens with the eggbeater. The magnitude of force

changes. The force required to turn the handle is greater than the force applied to the frosting by the blades. This results in a mechanical advantage of less than one.

TYPES OF GEARS

When two shafts are not lying in the same straight line, but are parallel, you can transmit motion from

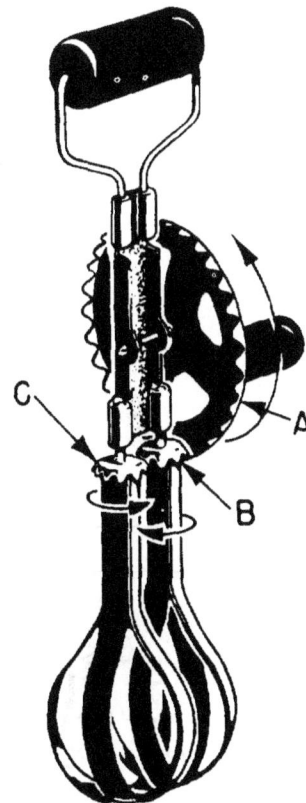

Figure 6-1.—A simple gear arrangement.

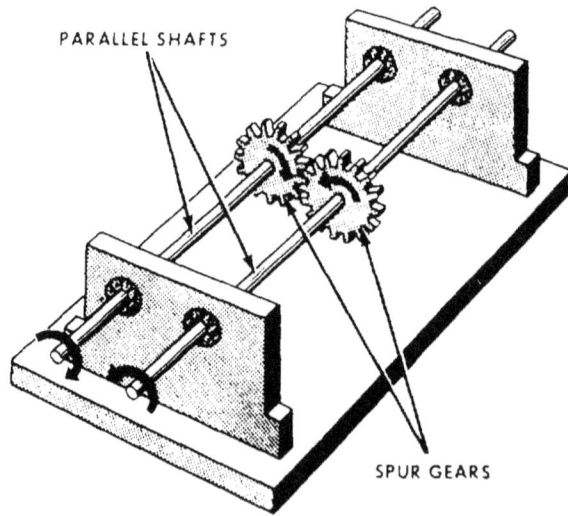

PARALLEL SHAFTS

SPUR GEARS

Figure 6-2.4-Spur gears coupling two parallel shafts.

one to the other by spur gears. This setup is shown in figure 6-2.

Spur gears are wheels with mating teeth cut in their surfaces so that one can turn the other without slippage. When the mating teeth are cut so that they are parallel to the axis of rotation, as shown in figure 6-2, the gears are called straight spur gears.

When two gears of unequal size are meshed together, the smaller of the two is usually called a pinion. By unequal size, we mean an unequal number of teeth causing one gear to be a larger diameter than the other. The teeth, themselves, must be of the same size to mesh properly.

The most commonly used gears are the straight spur gears. Often you'll run across another type of spur gear called the helical spur gear.

In helical gears the teeth are cut slantwise across the working face of the gear. One end of the tooth, therefore, lies ahead of the other. Thus, each tooth has a leading end and a trailing end. Figure 6-3, view A, shows you the construction of these gears.

In the straight spur gears, the whole width of the teeth comes in contact at the same time. However, with helical (spiral) gears, contact between two teeth starts first at the leading ends and moves progressively across the gear faces until the trailing ends are in contact. This kind of meshing action keeps the gears in constant contact with one another. Therefore, less lost motion and smoother, quieter action is possible. One disadvantage of this helical spur gear is the tendency of each gear to thrust or push axially on its shaft. It is necessary to put a special thrust bearing at the end of the shaft to counteract this thrust.

You do not need thrust bearings if you use herringbone gears like those shown in figure 6-4. Since the teeth on each half of the gear are cut in opposite directions, each half of the gear develops a thrust that counterbalances the other half. You'll find herringbone gears used mostly on heavy machinery.

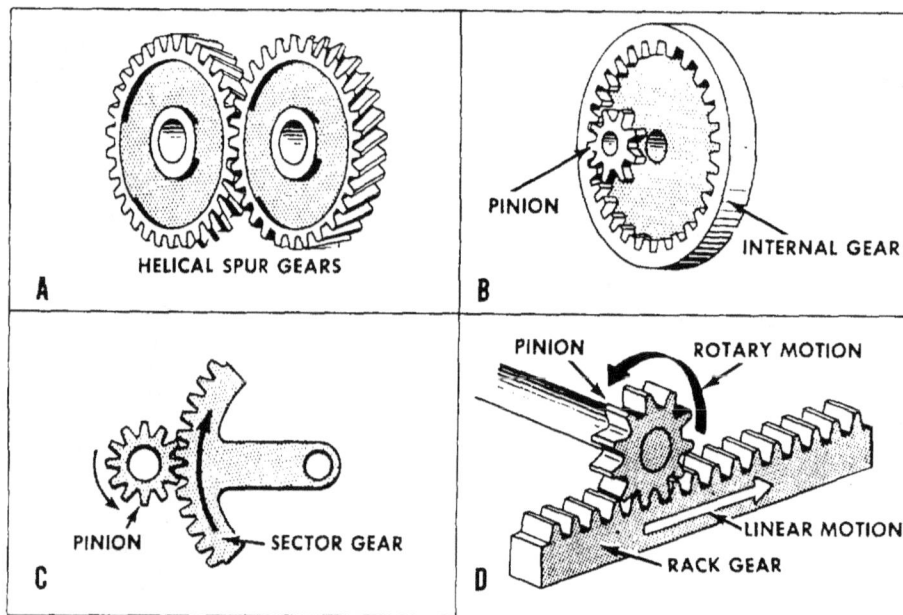

A HELICAL SPUR GEARS

B PINION INTERNAL GEAR

C PINION SECTOR GEAR

D PINION ROTARY MOTION LINEAR MOTION RACK GEAR

Figure 6-3.-Gear types.

Figure 6-4.—Herringbone gear.

Figure 6-3, views B, C, and D, also shows you three other gear arrangements in common use.

The internal gear in figure 6-3, view B, has teeth on the inside of a ring, pointing inward toward the axis of rotation. An internal gear is meshed with an external gear, or pinion, whose center is offset from the center of the internal gear. Either the internal or pinion gear can be the driver gear, and the gear ratio is calculated the same as for other gears—by counting teeth.

You only need a portion of a gear where the motion of the pinion is limited. You use the sector gear shown in figure 6-3, view C, to save space and material. The rack and pinion in figure 6-3, view D, are both spur gears. The rack is a piece cut from a gear with an extremely large radius. The rack-and-pinion arrangement is useful in changing rotary motion into linear motion.

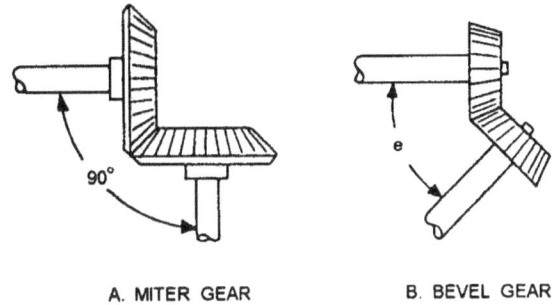

Figure 6-5.-Bevel gears.

THE BEVEL GEAR

So far most of the gears you've learned about transmit motion between parallel shafts. However, when shafts are not parallel (at an angle), we use another type of gear called the bevel gear. This type of gear can connect shafts lying at any given angle because you can bevel them to suit the angle.

Figure 6-5, view A, shows a special case of the bevel gear-the miter gear. You use the miter gears to connect shafts having a 90-degree angle; that means the gear faces are beveled at a 45-degree angle.

You can see in figure 6-5, view B, how bevel gears are designed to join shafts at any angle. Gears cut at any angle other than 45 degrees are bevel gears.

The gears shown in figure 6-5 are straight bevel gears, because the whole width of each tooth comes in contact with the mating tooth at the same time. However, you'll run across spiral bevel gears with teeth cut to have advanced and trailing ends. Figure 6-6 shows you what spiral bevel gears look like. They have the same advantage as other spiral (helical) gears—less lost motion and smoother, quieter operation.

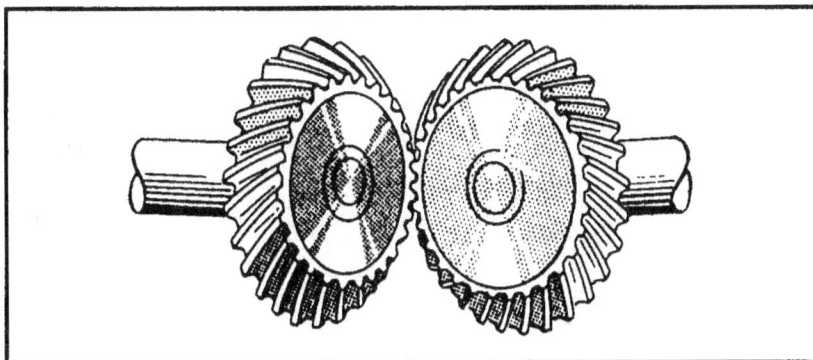

Figure 6-6.-Spiral bevel gears.

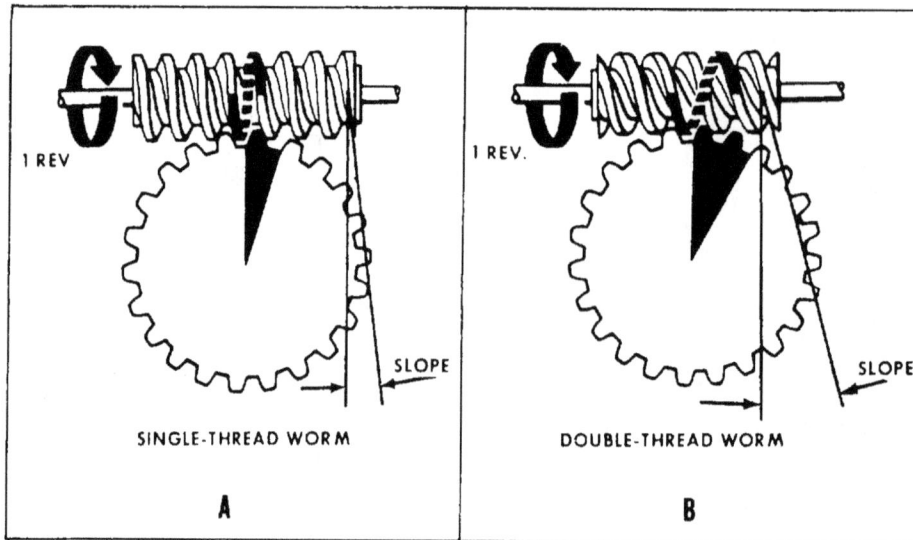

Figure 6-7.—Worm gears.

THE WORM AND WORM WHEEL

Worm and worm-wheel combinations, like those in figure 6-7, have many uses and advantages. However, it's better to understand their operating theory before learning of their uses and advantages.

Figure 6-7, view A, shows the action of a single-thread worm. For each revolution of the worm, the worm wheel turns one tooth. Thus, if the worm wheel has 25 teeth, the gear ratio is 25:1.

Figure 6-7, view B, shows a double-thread worm. For each revolution of the worm in this case, the worm wheel turns two teeth. That makes the gear ratio 25:2 if the worm wheel has 25 teeth.

A triple-thread worm would turn the worm wheel three teeth per revolution of the worm.

A worm gear is a combination of a screw and a spur gear. You can obtain remarkable mechanical advantages with this arrangement. You can design worm drives so that only the worm is the driver-the spur cannot drive the worm. On a hoist, for example, you can raise or lower the load by pulling on the chain that turns the worm. If you let go of the chain, the load cannot drive the spur gear; therefore, it lets the load drop to the deck. This is a nonreversing worm drive.

GEARS USED TO CHANGE DIRECTION

The crankshaft in an automobile engine can turn in only one direction. If you want the car to go backwards, you must reverse the effect of the engine's rotation. This is done by a reversing gear in the transmission, not by reversing the direction in which the crankshaft turns.

A study of figure 6-8 will show you how gears are used to change the direction of motion. This is a schematic diagram of the sight mounts on a Navy gun. If you crank the range-adjusting handle (A) in a clockwise direction, the gear (B) directly above it will rotate in a counterclockwise direction. This motion causes the two pinions (C and D) on the shaft to turn in the same direction as the gear (B) against the teeth cut in the bottom of the table. The table is tipped in the direction indicated by the arrow.

As you turn the deflection-adjusting handle (E) in a clockwise direction, the gear (F) directly above it turns

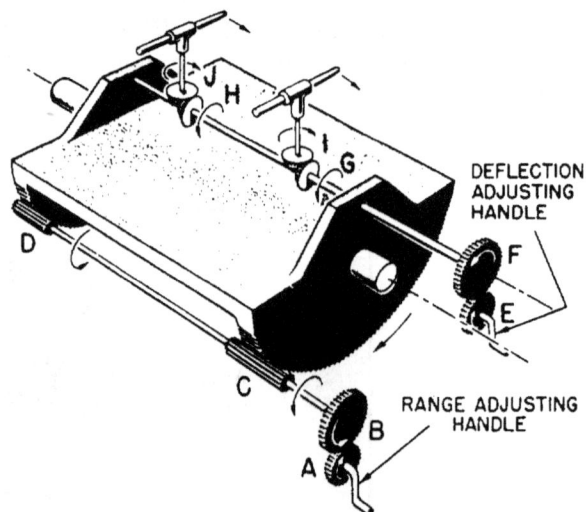

Figure 6-8.-Gears change direction of applied motion.

in the opposite direction. Since the two bevel gears (G and H) are fixed on the shaft with F, they also turn. These bevel gears, meshing with the horizontal bevel gears (I and J), cause I and J to swing the front ends of the telescopes to the right. Thus with a simple system of gears, it is possible to keep the two telescopes pointed at a moving target. In this and many other applications, gears serve one purpose: to change the direction of motion.

GEARS USED TO CHANGE SPEED

As you've already seen in the eggbeater, you use gears to change the speed of motion. Another example of this use of gears is in your clock or watch. The mainspring slowly unwinds and causes the hour hand to make one revolution in 12 hours. Through a series-or train-of gears, the minute hand makes one revolution each hour, while the second hand goes around once per minute.

Figure 6-9 will help you to understand how speed changes are possible. Wheel A has 10 teeth that mesh with the 40 teeth on wheel B. Wheel A will have to rotate four times to cause B to make one revolution. Wheel C is rigidly fixed on the same shaft with B. Thus, C makes the same number of revolutions as B. However, C has 20 teeth and meshes with wheel D, which has only 10 teeth. Hence, wheel D turns twice as fast as wheel C. Now, if you turn A at a speed of four revolutions per second, B will rotate at one revolution per second. Wheel C also moves at one revolution per second and causes D to turn at two revolutions per second. You get out two revolutions per second after having put in four revolutions per second. Thus, the overall speed reduction is 2/4—or 1/2—that means you got half the speed out of the last driven wheel you put into the first driver wheel.

You can solve any gear speed-reduction problem with this formula:

$$S_2 = S_1 \times \frac{T_1}{T_2},$$

where

S_1 = speed of first shaft in train

S_2 = speed of last shaft in train

T_1 = product of teeth on all drivers

T_2 = product of teeth on all driven gears

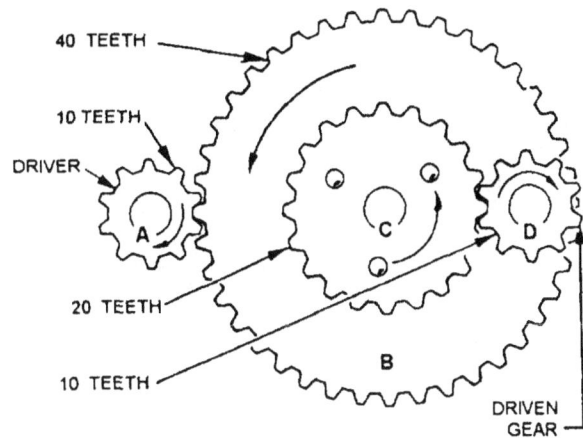

Figure 6-9.-Gears can change speed of applied motion.

Now use the formula on the gear train of figure 6-9.

$$S_2 = S_1 \times \frac{T_1}{T_2} = 4 \times \frac{10 \times 10}{40 \times 10}$$

$$= \frac{800}{400} = 2 \text{ revs. per sec.}$$

To obtain any increase or decrease in speed you, must choose the correct gears for the job. For example, the turbines on a ship have to turn at high speeds-say 5,800 rpm—if they are going to be efficient. The propellers, or screws, must turn rather slowly—say 195 rpm—to push the ship ahead with maximum efficiency. So, you place a set of reduction gears between the turbines and the propeller shaft.

When two external gears mesh, they rotate in opposite directions. Often you'll want to avoid this. Put a third gear, called an idler, between the driver and the driven gear. Don't let this extra gear confuse you on speeds. Just neglect the idler entirely. It doesn't change the gear ratio at all, and the formula still applies. The idler merely makes the driver and its driven gear turn in the same direction. Figure 6-10 shows you how this works.

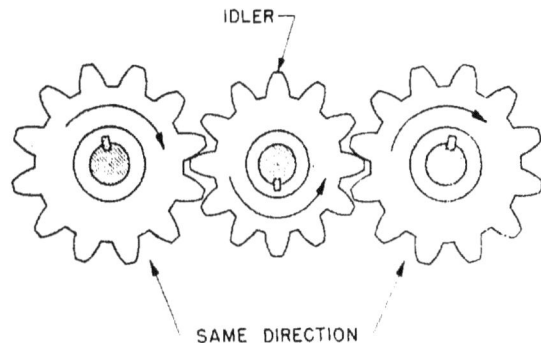

Figure 6-10.-An idler gear.

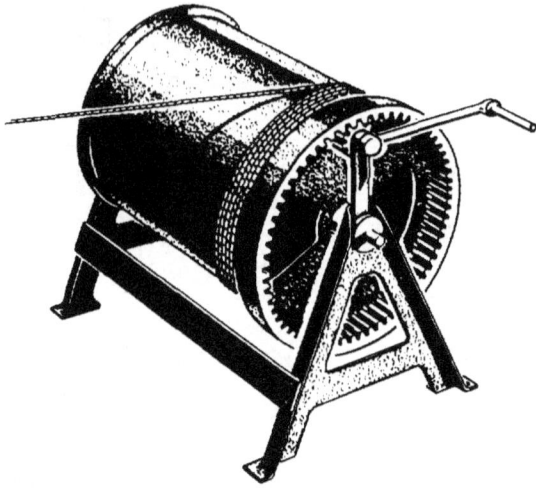

Figure 6-11.-Cable winch.

GEARS USED TO INCREASE MECHANICAL ADVANTAGE

We use gear trains to increase mechanical advantage. In fact, wherever there is a speed reduction, you multiply the effect of the effort. Look at the cable winch in figure 6-11. The crank arm is 30 inches long, and the drum on which the cable is wound has a 15-inch radius. The small pinion gear has 10 teeth, which mesh with the 60 teeth on the internal spur gear. You will find it easier to figure the mechanical advantage of this machine if you think of it as two machines.

First, figure out what the gear and pinion do for you. You find the theoretical mechanical advantage (T.M.A.) of any arrangement of two meshed gears by using the following formula:

$$\text{T.M.A.} = \frac{T_o}{T_a}$$

In which,

T_o = number of teeth on driven gear;

T_a = number of teeth on driver gear.

In this case,

T_o = 60 and T_a = 10.

Then,

$$\text{T.M.A.} = \frac{T_o}{T_a} = \frac{60}{10} = 6$$

Now, figure the mechanical advantage for the other part of the machine-a simple wheel-and-axle arrangement consisting of the crank arm and the drum. Divide the distance the effort moves ($2\pi R$) in making one complete revolution by the distance the cable is drawn up in one revolution of the drum ($2\pi r$).

$$\text{T.M.A.} = \frac{2\pi R}{2\pi r} = \frac{R}{r} = \frac{30}{15} = 2$$

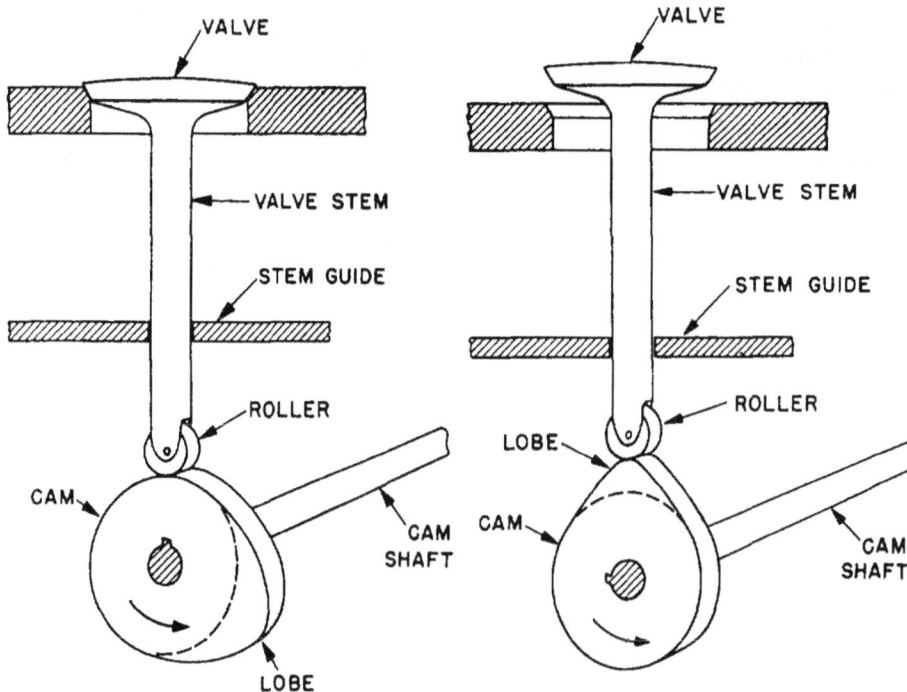

Figure 6-12.-Camdriven valve.

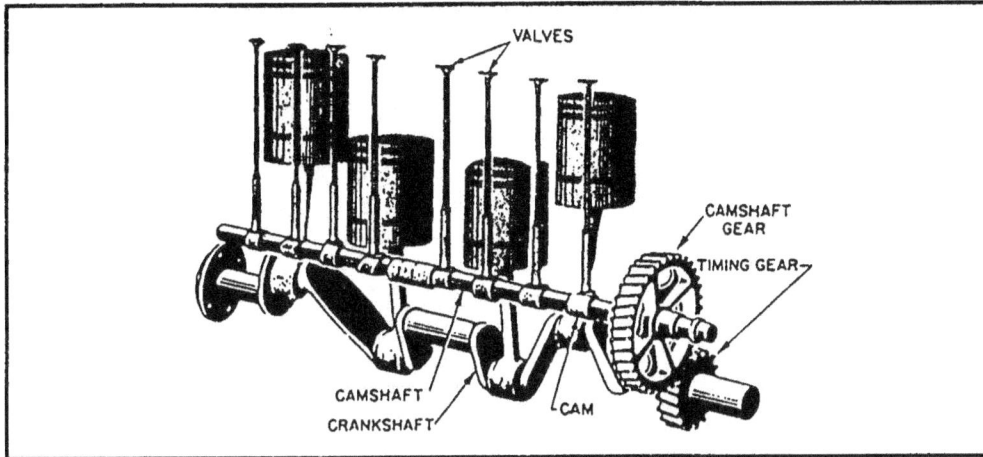

Figure 6-13.—Automobile valve gear.

The total, or overall, theoretical mechanical advantage of a compound machine is equal to the product of the mechanical advantages of the several simple machines that make it up. In this case you considered the winch as two machines—one having a mechanical advantage of 6 and the other a mechanical advantage of 2. Therefore, the overall theoretical mechanical advantage of the winch is 6 x 2, or 12. Since friction is always present, the actual mechanical advantage may be only 7 or 8. Even so, by applying a force of 100 pounds on the handle, you could lift a load of 700 to 800 pounds.

CAM

You use gears to produce circular motion. However, you often want to change rotary motion into up-and-down, or linear, motion. You can use cams to do this. For example, in figure 6-12 the gear turns the cam shaft. A cam is keyed to the shaft and turns with it. The design on the cam has an irregular shape that moves the valve stem up and down. It gives the valve a straight-line motion as the cam shaft rotates.

When the cam shaft rotates, the high point (lobe) of the cam raises the valve to its open position. As the shaft continues to rotate, the high point of the cam passes, lowering the valve to a closed position.

A set of cams, two to a cylinder, driven by timing gears from the crankshaft operate the exhaust and intake valves on the gasoline automobile engine as shown in figure 6-13. We use cams in machine tools and other devices to make rotating gears and shafts do up-and-down work.

ANCHOR WINCH

One of the gear systems you'll get to see frequently aboard ship is that on the anchor winch. Figure 6-14 shows you one type in which you can readily see how the wheels go around. The winch engine or motor turns the driving gear (A). This gear has 22 teeth, which mesh with the 88 teeth on the large wheel (B). Thus, you know that the large wheel makes one revolution for every four revolutions of the driving gear (A). You get a 4-to-1 theoretical mechanical advantage out of that pair. Secured to the same shaft with B is the small spur gear (C), covered up here. The gear (C) has 30 teeth that mesh with the 90 teeth on the large gear (D), also covered up.

Figure 6-14.—An anchor winch.

ship's wheel turns the small pinion (A). This pinion causes the internal spur gear to turn. Notice that this arrangement has a large mechanical advantage.

Now you see that when the center pinion (P) turns, it meshes with the two vertical racks. When the wheel turns full to the right, one rack moves downward and the other moves upward to the position of the racks. Attached to the bottom of the racks are two hydraulic pistons that control the steering of the ship. You'll get some information on this hydraulic system in a later chapter.

SUMMARY

These are the important points you should keep in mind about gears:

Gears can do a job for you by changing the direction, speed, or size of the force you apply.

When two external gears mesh, they always turn in opposite directions. You can make them turn in the same direction by placing an idler gear between the two.

The product of the number of teeth on each of the driver gears divided by the product of the number of teeth on each of the driven gears gives you the speed ratio of any gear train.

The theoretical mechanical advantage of any gear train is the product of the number of teeth on the driven gear wheels, divided by the product of the number of teeth on the driver gears.

The overall theoretical mechanical advantage of a We compound machine is equal to the product of the theoretical mechanical advantages of all the simple machines that make it up.

We can use cams to change rotary motion into linear motion.

Figure 6-15.—A steering mechanism.

The advantage from C to D is 3 to 1. The sprocket wheel to the far left, on the same shaft with D, is called a wildcat. The anchor chain is drawn up over this. Every second link is caught and held by the protruding teeth of the wildcat. The overall mechanical advantage of the winch is 4 x 3, or 12 to 1.

RACK AND PINION

Figure 6-15 shows you an application of the rack and pinion as a steering mechanism. Turning the

CHAPTER 7

WORK

CHAPTER LEARNING OBJECTIVES

Upon completion of this chapter, you should be able to do the following:

● *Define the term "work" when applied to mechanical power.*

MEASUREMENT

You know that machines help you to do work. What is work? Work doesn't mean simply applying a force. If that were so, you would have to consider that the sailor in figure 7-1 is doing work. He is busy applying his 220-pound force on the seabag. However, no work is being done!

Work in the mechanical sense, is done when a resistance is overcome by a force acting through a measurable distance. Now, if that sailor were to lift his 90-pound bag off the deck and put it on his bunk, he would be doing work. He would be overcoming a resistance by applying a force through a distance.

Notice that work involves two factors-force and movement through a distance. You measure force in pounds and distance in feet. Therefore, you measure work in units called foot-pounds. You do 1 foot-pound of work when you lift a 1-pound weight through a height

of 1 foot, You also do 1 foot-pound of work when you apply 1 pound of force on any object through a distance of 1 foot. Writing this as a formula, it becomes—

$$\underset{\text{(foot-pounds)}}{\text{WORK}} = \underset{\text{(pounds)}}{\text{FORCE}} = \underset{\text{(feet)}}{\text{DISTANCE}}$$

Thus, if you lift a 90-pound bag through a vertical distance of 5 feet, you will do

WORK = 90 X 5 = 450 ft-lb.

You should remember two points about work

1. In calculating the work done, you measure the actual resistance being overcome. The resistance is not necessarily the weight of the object you want to move. To understand this more clearly, look at the job the sailor in figure 7-2 is doing. He is pulling a 900-pound load of supplies 200 feet along the dock. Does this mean that he

Figure 7-1.—No work is being done.

Figure 7-2.—Working against friction.

Figure 7-3.—No motion, no work.

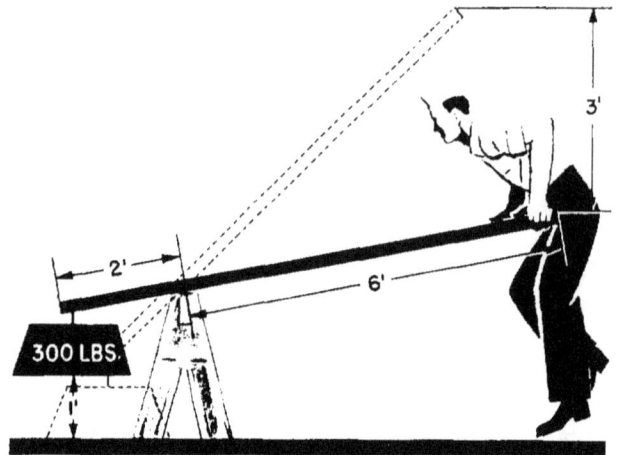

Figure 7-4.—Push'em up.

is doing 900 x 200, or 180,000 foot-pounds of work? Of course not. He isn't working against the pull of gravity-or the total weight—of the load. He's pulling only against the rolling friction of the truck and that may be as little as 90 pounds. That is the resistance that is being overcome. Always be sure you know what resistance is being overcome by the effort, as well as the distance through which it is moved. The resistance in one case may be the weight of the object; in another it may be the frictional resistance of the object as it is dragged or rolled along the deck.

2. You have to move the resistance to do any work on it. Look at the sailor in figure 7-3. The poor guy has been holding that suitcase for 15 minutes waiting for the bus. His arm is getting tired; but according to the definition of work, he isn't doing any because he isn't moving the suitcase. He is merely exerting a force against the pull of gravity on the bag.

You already know about the mechanical advantage of a lever. Now consider how it can be used to get work done easily. Look at figure 7-4. The load weighs 300 pounds, and the sailor wants to lift it up onto a platform a foot above the deck. How much work must he do? Since he must raise 300 pounds 1 foot, he must do 300 x 1, or 300 foot-pounds of work.

He can't make this weight any smaller with any machine. If he uses the 8-foot plank as shown, he can do the amount of work by applying a smaller force through a longer distance. Notice that he has a mechanical advantage of 3, so a 100-pound push down on the end of the plank will raise the 300-pound crate. Through how long a distance will he have to exert that 100-pound push? If he neglects friction, the work he exerts on the machine will be equal to the work done by the machine. In other words,

work put in = work put out.

Since Work = Force x Distance, you can substitute Force x Distance on each side of the work equation. Thus:

$$F_1 \times S_1 = F_2 \times S_2$$

in which

F_1 = effort applied, in pounds

S_1 = distance through which effort moves, in feet

F_2 = resistance overcome, in pounds

S_2 = distance resistance is moved, in feet

Now substitute the known values, and you get:

$$100 \times S_1 = 300 \times 1$$

$$S_1 = 3 \text{ feet}$$

The advantage of using the lever is not that it makes any less work for you, but it allows you to do the job with the force at your command. You'd probably have some difficulty lifting 300 pounds directly upward without a machine to help you!

Figure 7-5.—A block and tackle makes work easier.

Figure 7-6.—A big push.

A block and tackle also makes work easier. Like any other machine, it can't decrease the total amount of work to be done. With a rig like the one shown in figure 7-5, the sailor has a mechanical advantage of 5, neglecting friction. Notice that five parts of the rope go to and from the movable block. To raise the 600-pound load 20 feet, he needs to exert a pull of only one-fifth of 600—or 120 pounds. He is going to have to pull more than 20 feet of rope through his hands to do this. Use the formula again to figure why this is so:

Work input = work output

$$F_1 \times S_1 = F_2 \times S_2$$

And by substituting the known values:

$$120 \times S_1 = 600 \times 20$$

$$S_1 = 100 \text{ feet.}$$

This means that he has to pull 100 feet of rope through his hands to raise the load 20 feet. Again, the advantage lies in the fact that a small force operating through a large distance can move a big load through a small distance.

The sailor busy with the big piece of machinery in figure 7-6 has his work cut out for him. He is trying to seat the machine squarely on its foundations. He must shove the rear end over one-half foot against a frictional resistance of 1,500 pounds. The amount of work to be done is 1,500 x 1/2, or 750 foot-pounds. He will have to apply at least this much force on the jack he is using. If the jack has a 2 1/2-foot handle— R = 2 1/2 feet—and the pitch of the jack screw is one-fourth inch, he can do the job with little effort. Neglecting friction, you can figure it out this way:

Work input = work output

$$F_1 \times S_1 = F_2 \times S_2$$

In which

F_1 = force in pounds applied on the handle;

S_1 = distance in feet that the end of the handle travels in one revolution;

F_2 = resistance to overcome;

S_2 = distance in feet that the head of the jack advanced by one revolution of the screw, or, the pitch of the screw.

And, by substitution,

$$F_1 \text{ x } 2 \text{ x } 3.14 \text{ x } 2 1/2 = 1,500 \text{ x } 1/48$$

since

$$1/4 \text{ inch } = 1/48 \text{ of a foot}$$

$$F_1 \text{ x } 2 \text{ x } 2 1/2 = 1,5000 \text{ x } 1/48$$

$$F_1 = 2 \text{ pounds}$$

The jack makes it theoretically possible for the sailor to exert a 1,500-pound push with a 2-pound effort. Look at the distance through which he must apply that effort. One complete turn of the handle represents a distance of 15.7 feet. That 15.7-foot rotation advances the piece of machinery only one-fourth of an inch, or one-forty-eighth of a foot. You gain force at the expense of distance.

FRICTION

Suppose you are going to push a 400-pound crate up a 12-foot plank; the upper end is 3 feet higher than the lower end. You decide that a 100-pound push will do the job. The height you will raise the crate is one-fourth of the distance through which you will exert your push. The theoretical mechanical advantage is 4. Then you push on the crate, applying 100 pounds of force; but nothing happens! You've forgotten about the friction between the surface of the crate and the surface of the plank. This friction acts as a resistance to the movement of the crate; you must overcome this resistance to move the crate. In fact, you might have to push as much as 150 pounds to move it. You would use 50 pounds to overcome the frictional resistance, and the remaining 100 pounds would be the useful push that would move the crate up the plank.

Friction is the resistance that one surface offers to its movement over another surface. The amount of friction depends upon the nature of the two surfaces and the forces that hold them together.

In many instances fiction is useful to you. Friction helps you hold back the crate from sliding down the inclined ramp. The cinders you throw under the wheels of your car when it's slipping on an icy pavement increase the friction. You wear rubber-soled shoes in the gym to keep from slipping. Locomotives carry a supply of sand to drop on the tracks in front of the driving wheels to increase the friction between the wheels and the track. Nails hold structures together because of the friction between the nails and the lumber.

You make friction work for you when you slow up an object in motion, when you want traction, and when you prevent motion from taking place. When you want a machine to run smoothly and at high efficiency, you eliminate as much friction as possible by oiling and greasing bearings and honing and smoothing rubbing surfaces.

Where you apply force to cause motion, friction makes the actual mechanical advantage fall short of the theoretical mechanical advantage. Because of friction, you have to make a greater effort to overcome the resistance that you want to move. If you place a marble and a lump of sugar on a table and give each an equal push, the marble will move farther. That is because rolling friction is always less than sliding friction. You take advantage of this fact whenever you use ball bearings or roller bearings. See figure 7-7.

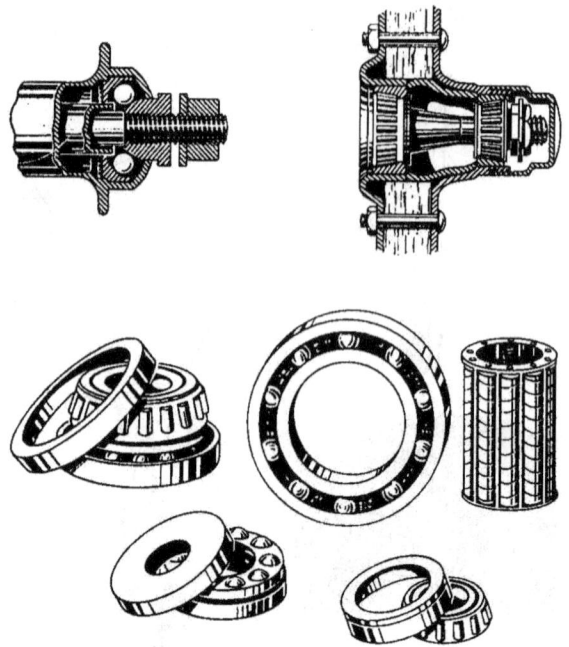

Figure 7-7.—These reduce friction.

Figure 7-8.—It saves wear and tear.

The Navy takes advantage of that fact that rolling friction is always less than sliding friction. Look at figure 7-8. This roller chock cuts down the wear and tear on lines and cables that are run through it. It also reduces friction and reduces the load the winch has to work against.

Figure 7-9.—Roller bitt saves line.

The roller bitt in figure 7-9 is another example of how you can cut down the wear and tear on lines or cable and reduce your frictional loss.

When you need one surface to move over another, you can decrease the friction with lubricants such as oil, grease, or soap. You can use a lubricant on flat surfaces and gun slides as well as on ball and roller bearings. A lubricant reduces frictional resistance and cuts down wear.

In many situations friction is helpful. However, many sailors have found out about this the hard way—on a wet, slippery deck. You'll find rough grain coverings are used on some of our ships. Here you have friction working for you. It helps you to keep your footing.

EFFICIENCY

To make it easier to explain machine operations, we have neglected the effect of friction on machines up to this point. Friction happens every time two surfaces move against one another. The work used in overcoming the frictional resistance does not appear in the work output. Therefore, it's obvious that you have to put more work into a machine than you get out of it. Thus, no machine is 100 percent efficient.

Take the jack in figure 7-6, for example. The chances are good that a 2-pound force exerted on the handle wouldn't do the job at all. You would need a pull of at least 10 pounds. This shows that only 2 pounds out of the 10 pounds, or 20 percent of the effort, is employed to do the job. The remaining 8 pounds of effort was is in overcoming the friction in the jack. Thus, the jack has an efficiency of only 20 percent. Most jacks are inefficient. However, even with this inefficiency, it is possible to deliver a huge push with a small amount of effort.

A simple way to calculate the efficiency of a machine is to divide the output by the input and convert it to a percentage:

$$\text{Efficiency} = \frac{\text{Output}}{\text{Input}}$$

Now go back to the block-and-tackle problem illustrated in figure 7-5. It's likely that instead of being able to lift the load with a 120-pound pull, the sailor would have to use a 160-pound pull through the 100 feet. You can calculate the efficiency of the rig by the following method:

$$\text{Efficiency} = \frac{\text{Output}}{\text{Input}} = \frac{F_2 \times S_2}{F_1 \times S_1}$$

and, by substitution,

$$\text{Efficiency} = \frac{600 \times 20}{160 \times 100} = 0.75 \text{ Or } 75 \text{ percent.}$$

Theoretically, with the mechanical advantage of 12 developed by the cable winch in figure 6-11, you can lift a 600-pound load with a 50-pound push on the handle. If the machine has an efficiency of 60 percent, how big a push would you actually have to apply? Actually, 50 + 0.60 = 83.3 pounds. You can check this yourself in the following manner:

$$\text{Efficiency} = \frac{\text{Output}}{\text{Input}}$$

$$= \frac{F_2 \times S_2}{F_1 \times S_1}$$

One revolution of the drum would raise the 600-pound load a distance S_2 of $2\pi r$, or 7.85 feet. To make the drum revolve once, the pinion gear must rotate six times by the handle, and the handle must turn through a distance S_1 of 6 x $2\pi R$, or 94.2 feet. Then, by substitution:

$$0.60 = \frac{600 \times 7.85}{F_1 \text{ X } 94.2}$$

$$F_1 = \frac{600 \times 7.85}{94.2 \times 0.60} = 83.3 \text{ pounds.}$$

Because this machine is only 60-percent efficient, you have to put 94.2 x 83.3, or 7,847 foot-pounds, of work into it to get 4,710 foot-pounds of work out of it. The difference (7,847 − 4,710 = 3,137 foot-pounds) is used to overcome friction within the machine.

SUMMARY

Here are some of the important points you should remember about friction, work and efficiency:

You do work when you apply a force against a resistance and move the resistance.

Since force is measured in pounds and distance is measured in feet, we measure work in foot-pounds. One foot-pound of work is the result of a 1-pound force, acting against a resistance through a distance of 1 foot.

Machines help you to do work by making it possible to move a large resistance through a small distance by the application of a small force through a large distance.

Since friction is present in all machines, more work must be done on the machine than the machine actually does on the load.

You can find the efficiency of any machine by dividing the output by the input.

Friction is the resistance that one surface offers to movement over a second surface.

Friction between two surfaces depends upon the nature of the materials and the size of the forces pushing them together.

CHAPTER 8

POWER

CHAPTER LEARNING OBJECTIVES

Upon completion of this chapter, you should be able to do the following:

- *Define the term "power."*

- *Determine horsepower ratings.*

It's all very well to talk about how much work a person can do. The payoff is how long it takes him or her to do it. Look at the sailor in figure 8-1. He has lugged 3 tons of bricks up to the second deck of the new barracks. However, it has taken him three 10-hour days—1,800 minutes-to do the job. In raising the 6,000 pounds 15 feet, he did 90,000 foot-pounds (ft-lb) of work. Remember, force x distance = work. Since it took him 1,800 minutes, he has been working at 90,000 ÷ 1,800, or 50 foot-pounds of work per minute.

That's power—the rate of doing work. Thus, power always includes a time element. Doubtless you could do the same amount of work in one 10-hour day, or 600 minutes. This would mean that you would work at the rate of 90,000 ÷ 600 = 150 foot-pounds per minute. You then would have a power value three times as much as that of the sailor in figure 8-1.

Apply the following formula:

$$\text{Power} = \frac{\text{Work, in ft-lb}}{\text{Time, in minutes}}$$

Figure 8-1.-Get a horse.

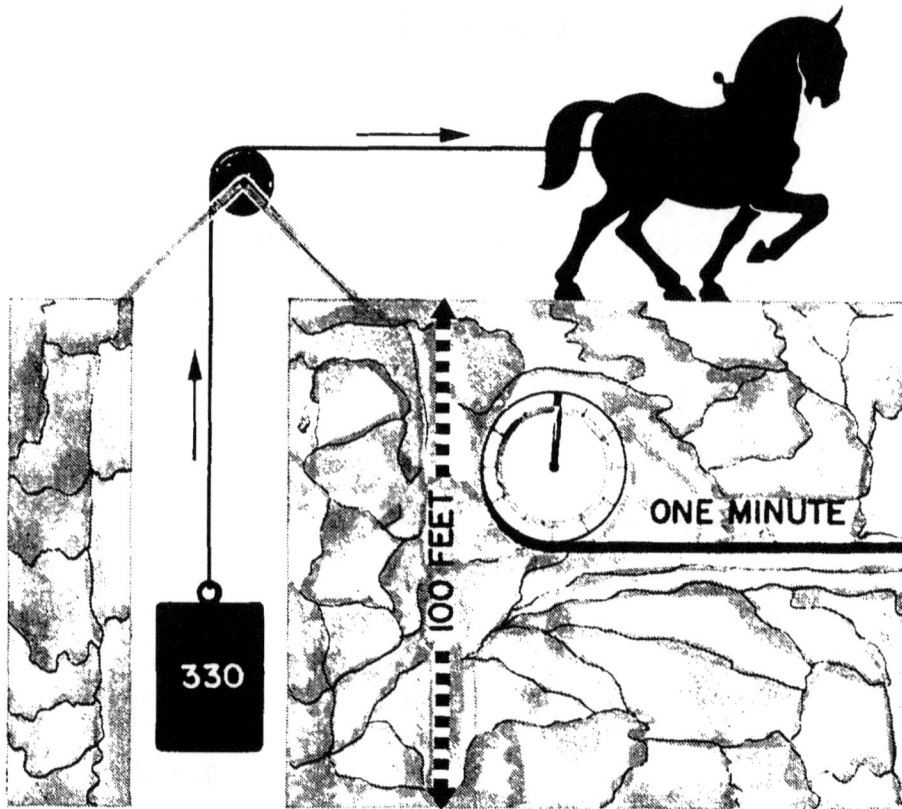

Figure 8-2.-One horsepower.

HORSEPOWER

You measure force in pounds, distance in feet, and work in foot-pounds. What is the common unit used for measuring power? It is called horsepower (hp). If you want to tell someone how powerful an engine is, you could say that it is many times more powerful than a man or an ox or a horse. But what man? and whose ox or horse? James Watt, the man who invented the steam engine, compared his early models with the horse. By experiment, he found that an average horse, hitched to a rig as shown in figure 8-2, could lift a 330-pound load straight up a distance of 100 feet in 1 minute. Scientists agree that 1 horsepower equals 33,000 foot-pounds of work done in 1 minute.

Since 60 seconds equals a minute, 1 horsepower is equal to $^{33,000}/_{60}$ = 550 foot-pounds per second. Use the following formula to figure horespower:

$$\text{Horsepower} = \frac{\text{Power (in ft–lb per min)}}{33,000}$$

CALCULATING POWER

It isn't difficult to figure how much power you need to do a certain job in a given length of time. Nor is it difficult to predict what size engine or motor you need to do it. Suppose an anchor winch must raise a 6,600-pound anchor through 120 feet in 2 minutes. What must be the theoretical horsepower rating of the motor on the winch?

The first step is to find the rate at which the work must be done using the formula:

$$\text{Power} = \frac{\text{work}}{\text{time}} = \frac{\text{force} \times \text{distance}}{\text{time}}$$

Substitute the known values in the formula, and you get:

$$\text{Power} = \frac{6,600 \times 120}{2} = 396,000 \text{ ft–lb/min}$$

So far, you know that the winch must work at a rate of 396,000 ft-lb/min. To change this rate to horsepower, you divide by the rate at which the average horse can work—33,000 ft-lb/min.

$$\text{Horsepower} = \frac{\text{Power (in ft–lb per min)}}{33,000}$$

$$= \frac{396,000}{33,000} = 12 \text{ hp}$$

Theoretically, the winch would have to work at a rate of 12 horsepower to raise the anchor in 2 minutes. Of course, you've left out all friction in this problem, so the winch motor would actually have to be larger than 12 hp.

You raise planes from the hangar deck to the flight deck of a carrier on an elevator. Some place along the line, an engineer had to figure out how powerful the motor had to be to raise the elevator. It's not too tough when you know how. Allow a weight of 10 tons for the elevator and 5 tons for the plane. Suppose that you want to raise the elevator and plane 25 feet in 10 seconds and that the overall efficiency of the elevator mechanism is 70 percent. With that information you can figure what the delivery horsepower of the motor must be. Set up the formulas:

$$\text{Power} = \frac{\text{force} \times \text{distance}}{\text{time}}$$

$$\text{hp} = \frac{\text{power}}{33,000}$$

Substitute the known values in their proper places, and you have:

$$\text{Power} = \frac{30,000 \times 25 \text{ ft}}{10/60 \text{ minute}} = 4,500,000 \text{ ft–lb/min.}$$

$$\text{hp} = \frac{4,500,000}{33,000} = 136.4 \text{ hp}$$

So, you need 136.4 horsepower if the engine has 100 percent overall efficiency. You want to use 70 percent efficiency, so you use the formula:

$$\text{Efficiency} = \frac{\text{Output}}{\text{Input}}$$

$$\text{Input} = \frac{136.4}{0.70} = 194.8 \text{ hp}$$

This is the rate at which the engine must be able to work. To be on the safe side, you'd probably select a 200-horsepower auxiliary to do the job.

FIGURING THE HORSEPOWER RATING OF A MOTOR

You have probably seen the horsepower rating plates on electric motors. You may use several methods to determine this rating. One way to find the rating of a

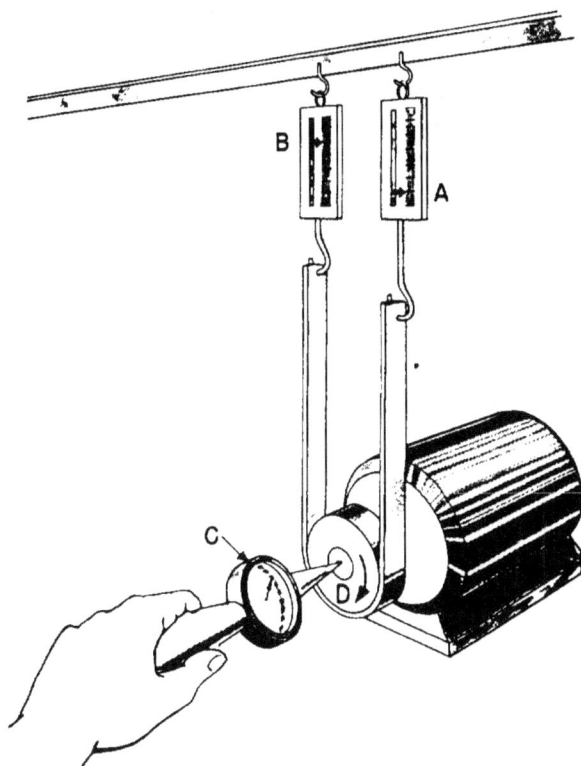

Figure 8-3.-A prony brake.

motor or a steam or gas engine is with the use of the prony brake. Figure 8-3 shows you the prony brake setup. A pulley wheel is attached to the shaft of the motor and a leather belt is held firmly against the pulley. Attached to the two ends of the belts are spring scales. When the motor is standing still, each scale reads the same— 15 points. When the pulley turns in a clockwise direction, the friction between the belt and the pulley makes the belt try to move with the pulley. Therefore, the pull on scale A will be greater than 15 pounds, and the pull on scale B will be less than 15 pounds.

Suppose that scale A reads 25 pounds and scale B reads 5 pounds. That tells you the drag, or the force against which the motor is working, is 25 – 5 = 20 pounds. In this case the normal speed of the motor is 1,800 revolutions per minute (rpm) and the diameter of the pulley is 1 foot.

You can find the number of revolutions by holding the revolution counter (fig. 8-3, C) against the end of the shaft for 1 minute. This counter will record the number of turns the shaft makes per minute. The distance (D) that any point on the pulley travels in 1 minute is equal to the circumference of the pulley times the number of revolutions or 3.14 x 1 x 1,800 = 5,652 ft.

You know that the motor is exerting a force of 20 pounds through that distance. The work done in 1 minute is equal to the force times the distance, or work = F x D = 20 x 5,652 = 113,040 ft-lb/min. Change this to horsepower:

$$\frac{113,040}{33,000} = 3.43 \text{ hp}$$

Two common motor or engine ratings with which you are familiar are the 1/16-hp motor of an electric mixer and the 1/4-hp motor of a washing machine.

SUMMARY

Remember two important points about power:

Power is the rate at which work is done.

Horsepower is the unit of measurement by which power is equivalent to 33,000 foot-pounds of work per minute, or 550 foot-pounds per second.

CHAPTER 9

FORCE AND PRESSURE

CHAPTER LEARNING OBJECTIVES

Upon completion of this chapter, you should be able to do the following:

• *Explain the difference in force and pressure.*

• *Discuss the operation of force- and pressure-measuring devices.*

By this time you should have a pretty good idea of what force is. Now you will learn the difference between force and pressure and how force affects pressure.

FORCE

Force is the pull of gravity exerted on an object or an object's thrust of energy against friction. You apply a force on a machine; the machine, in turn, transmits a force to the load. However, other elements besides men and machines can also exert a force. For example, if you've been out in a sailboat, you know that the wind can exert a force. Further, after the waves have knocked you on your ear a couple of times, you have grasped the idea that water, too, can exert a force. Aboard ship, from reveille to taps you are almost constantly either exerting forces or resisting them.

MEASURING FORCE

Weight is a measurement of the force, or pull of gravity, on an object. You've had a lot of experience in measuring forces. At times, you have estimated or "guessed' the weight of a package you were going to mail by "hefting" it. However, to find its accurate weight, you would have put it on a force-measuring device known as a scale. Scales are of two types: spring and balanced.

Spring Scale

You can readily measure force with a spring scale. An Englishman named Hooke invented the spring scale. He discovered that hanging a 1-pound weight on a spring caused the spring to stretch a certain distance and that hanging a 2-pound weight on the spring caused it to stretch twice as far. By attaching a pointer to the spring and inserting the pointer through a face, he could mark points on the face to indicate various measurements in pounds and ounces.

We use this type of scale to measure the pull of gravity-the weight-of an object or the force of a pull exerted against friction, as shown in figure 9-1.

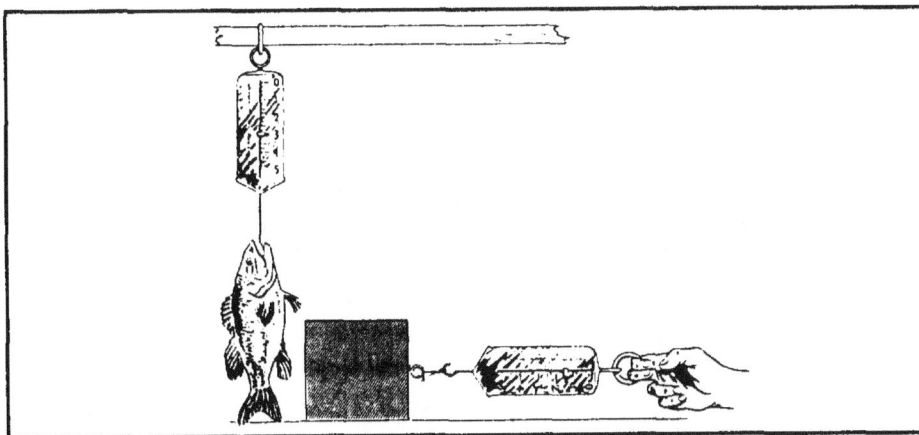

Figure 9-1.—You can measure force with a scale.

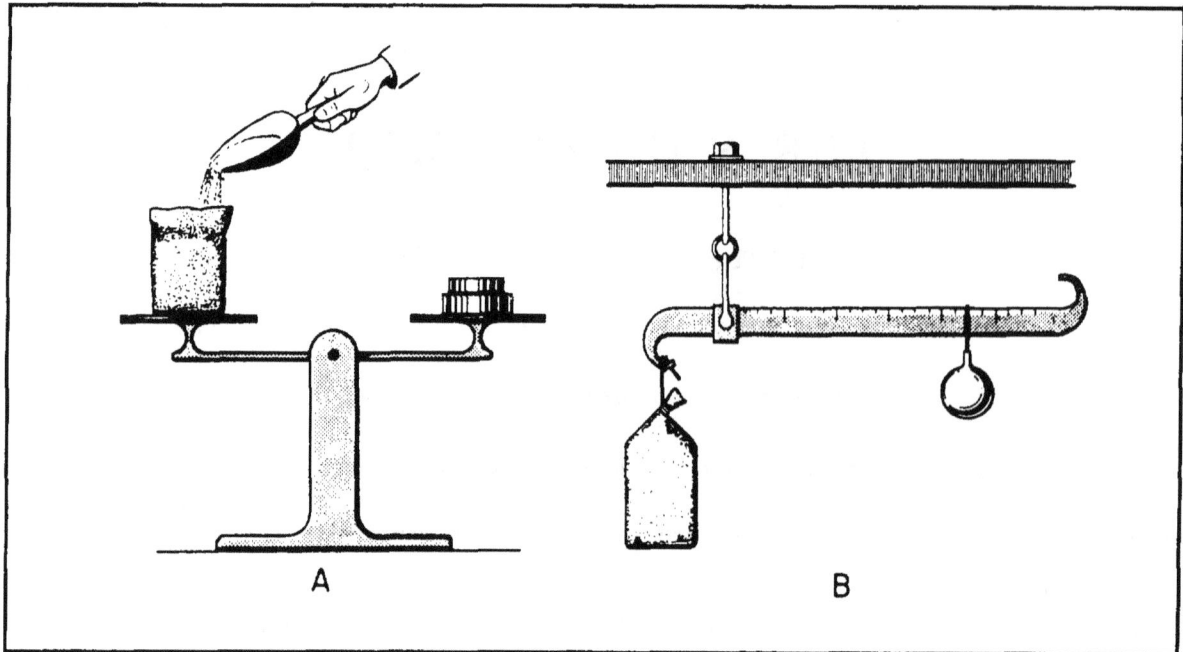

Figure 9-2.—Balances.

Unfortunately, the more springs are used, the more they lose their ability to snap back to their original position. Hence, an old spring or an overloaded spring will give inaccurate readings.

Balanced Scale

The problem with the spring-type scale eventually led to the invention of the balanced scale, shown in figure 9-2. This type of scale is an application of first-class levers. The one shown in figure 9-2, A, is the simplest type. Since the distance from the fulcrum to the center of each platform is equal, the scales balance when equal weights are placed on the platforms. With your knowledge of levers, you can figure out how the steel yard shown in figure 9-2, B, operates.

PRESSURE

Pressure is the amount of force within a specific area. You measure air, steam, and gas pressure and the fluid pressure in hydraulic systems in pounds per square inch (psi). However, you measure water pressure in pounds per square foot. You'll find more about pressure measurements in chapter 10. To help you better understand pressure, let's look at how pressure affects your ability to walk across snow.

Have you ever tried to walk on freshly fallen snow to have your feet break through the crust when you put your weight on it? If you had worn snowshoes, you could have walked across the snow without sinking; but do you know why? Snowshoes do not reduce your weight, or the amount of force, exerted on the snow; they merely distribute it over a larger area. In doing that, the snowshoes reduce the pressure per square inch of the force you exert.

Let's figure out how that works. If a man weighs 160 pounds, that weight, or force, is more or less evenly distributed by the soles of his shoes. The area of the soles of an average man's shoes is roughly 60 square inches. Each of those square inches has to carry $160 \div 60 = 2.6$ pounds of that man's weight. Since 2 to 6 pounds per square inch is too much weight for the snow crest to support, his feet break through.

When the man puts on snowshoes, he distributes his weight over an area of about 900 square inches, depending on the size of the snowshoes. The force on each of those square inches is equal to only $160 \div 900 = 0.18$ pounds. Therefore, with snowshoes on, he exerts a pressure of 0.18 psi. With this decreased pressure, the snow can easily support him.

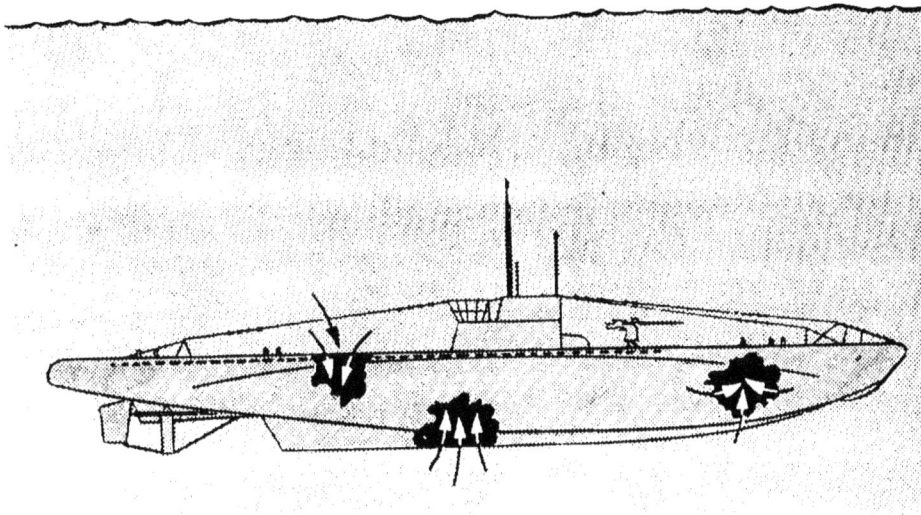

Figure 9-3.-Fluids exert pressure in all directions.

CALCULATING PRESSURE

To calculate pressure, divide the force by the area on which you apply force. Use the following formula:

$$\text{Pressure, in psi} = \frac{\text{Force, in lb}}{\text{Area, in sq in.}}$$

or

$$P = \frac{F}{A}$$

To understand this idea, follow this problem. A fresh water holding tank aboard a ship is 10 feet long, 6 feet wide, and 4 feet deep. Therefore, it holds 10 x 6 x 4, or 240, cubic feet of water. Each cubic foot of water weighs about 62.5 pounds. The total force outside the tank's bottom is equal to the weight of the water: 240 x 62.5, or 15,000 pounds. What is the pressure on the bottom of the tank? Since the weight is even on the bottom, you apply the formula $P = \frac{F}{A}$ and substitute the proper values for F and A. In this case, F= 15,000 pounds; the area of the bottom in square inches is 10 x 6 x 144, since 144 square inches = 1 square foot.

$$P = \frac{15,000}{1\,0\text{ x }6\text{ x }1\,4\,4}$$

Now work out the idea in reverse. You live at the bottom of the great sea of air that surrounds the earth. Because the air has weight—gravity pulls on the air too—the air exerts a force on every object that it surrounds. Near sea level that force on an area of 1 square inch is roughly 15 pounds. Thus, the air-pressure at sea level is about 15 psi. The pressure gets less and less as you go up to higher altitudes.

With your finger, mark out an area of 1 square foot on your chest. What is the total force pushing on your chest? Again use the formula $P = \frac{F}{A}$. Now substitute 15 psi for P and 144 square inches for A. Then, $F = 144$ x 15, or 2,160 pounds. The force on your chest is 2,160 pounds per square foot-more than a ton pushing against an area of 1 square foot. If no air were inside your chest to push outward with the same pressure, you'd be flatter than a bride's biscuit.

MEASURING FLUID PRESSURE

All fluids-both liquids and gases—exert pressure. A fluid at rest exerts equal pressure in all directions. As shown in figure 9-3, water will push through a hole in a submarine, whether it is in the top, the bottom, or in one of the sides.

Many jobs aboard ship will require you to know the pressure exerted by a gas or a liquid. For example, knowing the steam pressure inside a boiler is always important. You can use three different gauges to find the pressure of fluids: Bourdon gauge, Schrader gauge, and diaphragm gauge.

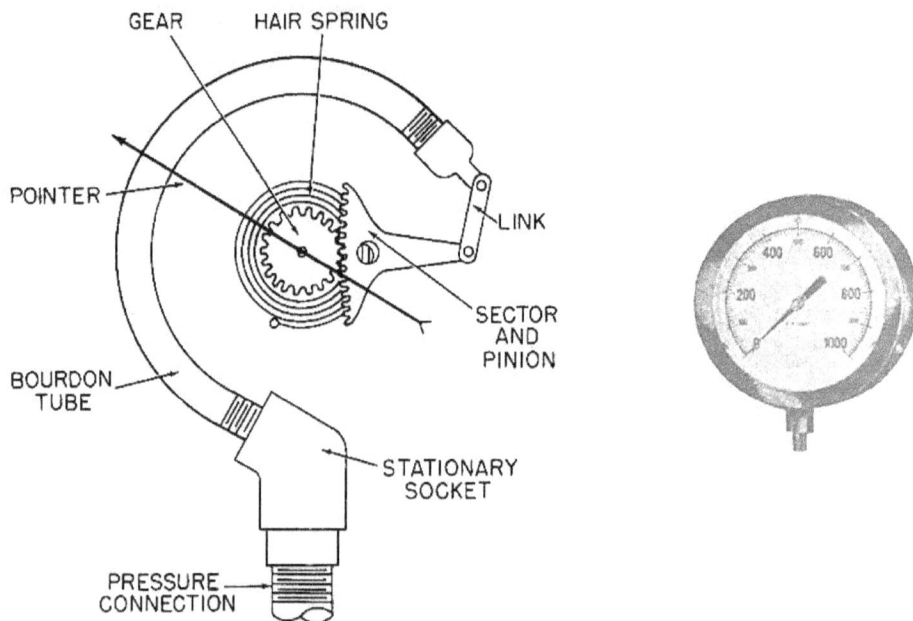

Figure 9-4.-The Bourdon gauge.

Bourdon Gauge

The Bourdon gauge is shown in figure 9-4. It works on the same principle as that of the snakelike, paper party whistle you get at a New Year party, which straightens when you blow into it.

Within the Bourdon gauge is a thin-walled metal tube, somewhat flattened and bent into the form of a C. Attached to its free end is a lever system that magnifies any motion of the free end of the tube. On the fixed end of the gauge is a fitting you thread into a boiler system. As pressure increases within the boiler, it travels through the tube. Like the snakelike paper whistle, the metal tube begins to straighten as the pressure increases inside of it. As the tube straightens, the pointer moves around a dial that indicates the pressure in psi.

The Bourdon gauge is a highly accurate but rather delicate instrument. You can easily damage it. In addition, it malfunctions if pressure varies rapidly. This problem was overcome by the development of another type of gauge, the Schrader. The Schrader gauge (fig. 9-5) is not as accurate as the Bourdon, but it is sturdy and suitable for ordinary hydraulic pressure measurements. It is especially suitable for fluctuating loads.

In the Schrader gauge, liquid pressure actuates a piston. The pressure moves up a cylinder against the resistance of a spring, carrying a bar or indicator with it over a calibrated scale. The operation of this

gauge eliminates the need for cams, gears, levers, and bearings.

Diaphragm Gauge

The diaphragm gauge gives sensitive and reliable indications of small pressure differences. We use the diaphragm gauge to measure the air pressure in the space between inner and outer boiler casings.

In this type of gauge, a diaphragm connects to a pointer through a metal spring and a simple linkage system (fig. 9-6). One side of the diaphragm is exposed to the pressure being measured, while the other side is exposed to the pressure of the atmosphere. Any increase in the pressure line moves the diaphragm upward against the spring, moving the pointer to a higher reading. When the pressure decreases, the spring moves the diaphragm downward, rotating the pointer to a lower reading. Thus, the position of the pointer is balanced between the pressure pushing the diaphragm upward and the spring action pushing down. When the gauge reads 0, the pressure in the line is equal to the outside air pressure.

MEASURING AIR PRESSURE

To the average person, the chief importance of weather is reference to it as an introduction to general conversation. At sea and in the air, advance knowledge of what the weather will do is a matter of great concern

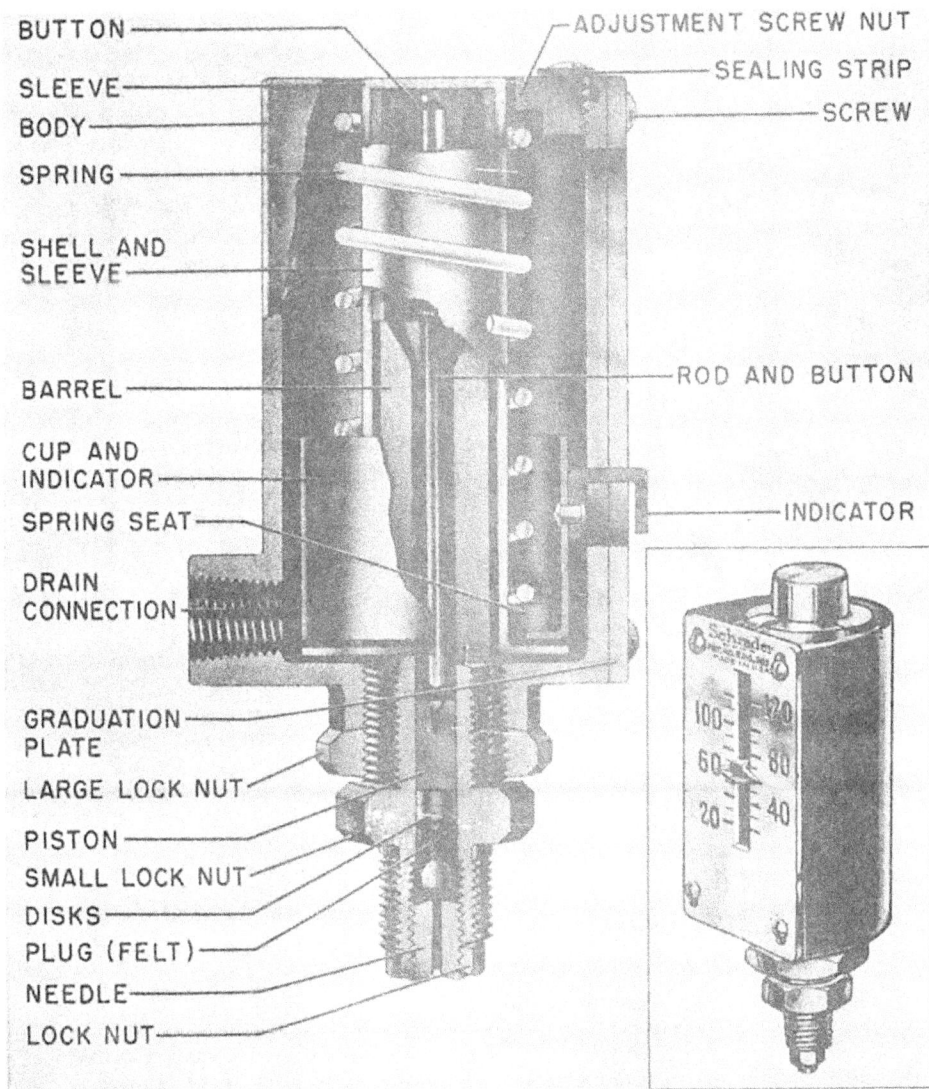

BUTTON
SLEEVE
BODY
SPRING
SHELL AND SLEEVE
BARREL
CUP AND INDICATOR
SPRING SEAT
DRAIN CONNECTION
GRADUATION PLATE
LARGE LOCK NUT
PISTON
SMALL LOCK NUT
DISKS
PLUG (FELT)
NEEDLE
LOCK NUT

ADJUSTMENT SCREW NUT
SEALING STRIP
SCREW
ROD AND BUTTON
INDICATOR

Figure 9-5.—The Schrader gauge.

to all hands. We plan or cancel operations on the basis of weather predictions. Accurate weather forecasts are made only after a great deal of information has been collected by many observers located over a wide area.

One of the instruments used in gathering weather data is the barometer, which measures air pressure. Remember, the air is pressing on you all the time. Normal atmospheric pressure is 14.7 psi. As the weather changes, the air pressure may be greater or less than normal. Air from high-pressure areas always moves toward low-pressure areas, and moving air—or wind-is one of the main causes of weather changes. In general, as air moves into a low-pressure area, it causes wind, rain, and storms. A high-pressure area usually enjoys clear weather. Ships use two types of barometers to measure air pressure: aneroid and mercurial.

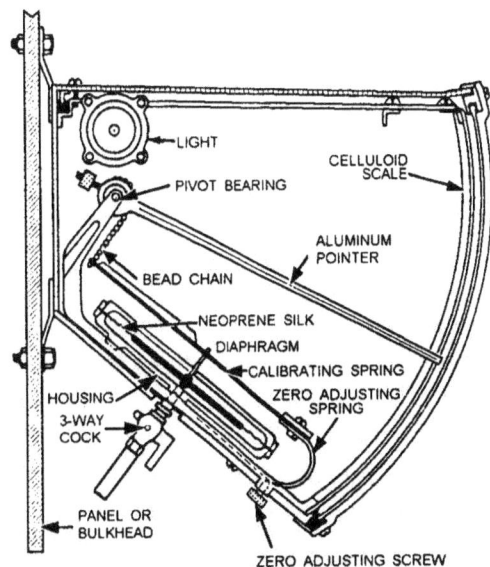

LIGHT
PIVOT BEARING
CELLULOID SCALE
ALUMINUM POINTER
BEAD CHAIN
NEOPRENE SILK DIAPHRAGM
CALIBRATING SPRING
ZERO ADJUSTING SPRING
HOUSING
3-WAY COCK
PANEL OR BULKHEAD
ZERO ADJUSTING SCREW

Figure 9-6.—Diaphragm pressure gauge.

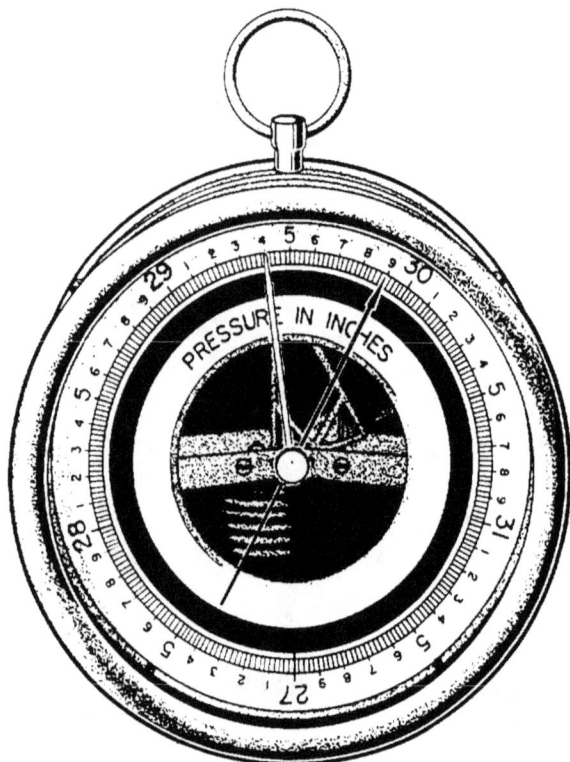

Figure 9-7.-An aneroid barometer.

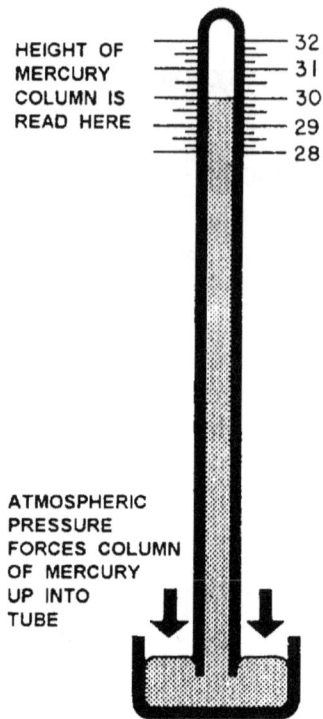

Figure 9-8.-A mercurial barometer.

Since air pressure affects weather, you can see why the use of a barometer is so important to ships. However, not so apparent is the importance of air pressure in the operation of the ship's engine. For that purpose air pressure is measured with a gauge called a manometer.

Aneroid Barometer

The aneroid barometer shown in figure 9-7 is an instrument that measures air pressure at sea level. It consists of a thin-walled metal box from which most of the air has been pumped and a dial indicating low- and high-pressure measurements. A pointer on the dial is connected to the box by a lever system. If the pressure of the atmosphere increases, it squeezes the sides of the box. This squeeze causes the pointer to move toward the high-pressure end of the dial. If the pressure decreases, the sides of the box expand outward. That causes the pointer to move toward the low-pressure end of the dial. Notice that the numbers on the dial are from 27 to 31. This scale of numbers is used because average sea level pressure is 29.92 inches and readings below 27 inches or above 31 inches are rarely seen.

Mercurial Barometer

Figure 9-8 illustrates a mercurial barometer. It consists of a glass tube on which measurements are indicated; the tube is partially filled with mercury. The upper end, which is closed, contains a vacuum above the mercury. The lower end, which is open, is submerged in a cup of mercury that is open to the atmosphere. The atmosphere presses down on the mercury in the cup and pushes the mercury up in the tube. The greater the air pressure, the higher the rise of mercury within the tube. At sea level, the normal pressure is 14.7 psi, and the height of the mercury in the tube is 30 inches. As the air pressure increases or decreases from day to day, the height of the mercury rises or falls. A mercury barometer aboard ship mounts in gimbals to keep it in a vertical position despite the rolling and pitching of the ship.

The dial of most gauges indicate relative pressure; that is, it is either greater or less than normal. Remember-the dial of an aneroid barometer always indicates absolute pressure, not relative. When the pressure exerted by any gas is less than 14.7 psi, you have what we call a partial vacuum.

Manometer

The condensers on steam turbines operate at a pressure well below 14.7 psi. Steam under high pressure

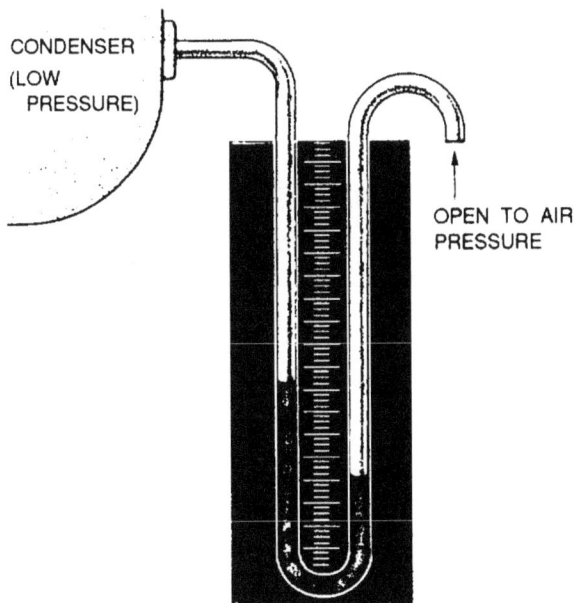

CONDENSER
(LOW
    PRESSURE)

OPEN TO AIR
PRESSURE

Figure 9-9.-A manometer.

runs into the turbine and causes the rotor to turn. After it has passed through the turbine, it still exerts a back pressure against the blades. If the back pressure were not reduced, it would build until it became as great as that of the incoming steam and prevent the turbine from turning at all. Therefore, the exhaust steam is run through pipes surrounded by cold sea water to reduce the back pressure as much as possible. The cold temperature causes the steam in the pipes to condense into water, and the pressure drops well below atmospheric pressure.

The engineer needs to know the pressure in the condensers at all times. To measure this reduced pressure, or partial vacuum, the engineer uses a gauge called a manometer. As shown in figure 9-9, it consists of a U-shaped tube. One end is connected to the low-pressure condenser, and the other end is open to the air. The tube is partially filled with colored water. The normal air pressure against the colored water is greater than the low pressure of the steam from the condenser. Therefore, the colored water is forced part of the way into the left arm of the tube. A scale between the two arms of the U indicates the difference in the height of the two columns of water. This difference tells the engineer the degree of vacuum-or how much below atmospheric pressure the pressure within the condenser is.

SUMMARY

You should remember seven points about force and pressure:

A force is a push or a pull exerted on or by an object.

You measure force in pounds.

Pressure is the force per unit area exerted on an object or exerted by an object. You measure it in pounds per square inch (psi).

You calculate pressure by the formula $P = \dfrac{F}{A}$.

Spring scales and lever balances are familiar instruments you use for measuring forces. Bourdon gauges, barometers, and manometers are instruments for the measurement of pressure.

The normal pressure of the air is 14.7 psi at sea level.

Pressure is generally relative; that is, it is sometimes greater—sometimes less—than normal air pressure. Pressure that is less than the normal air pressure is called a vacuum.

CHAPTER 10

HYDROSTATIC AND HYDRAULIC MACHINES

CHAPTER LEARNING OBJECTIVES

Upon completion of this chapter, you will be able to do the following:

- *Explain the difference between hydrostatic and hydraulic liquids.*

- *Discuss the uses of hydrostatic machines.*

- *Discuss the uses of hydraulic machines.*

In this chapter we will discuss briefly the pressure of liquids: (1) hydrostatic (liquids at rest) and (2) hydraulic (liquids in motion). We will discuss the operation of hydrostatic and hydraulic machines and give applications for both types.

HYDROSTATIC PRESSURE

You know that liquids exert pressure. The pressure exerted by seawater, or by any liquid at rest, is known as hydrostatic pressure.

If you are billeted on a submarine, you are more conscious of the hydrostatic pressure of seawater. When submerged, your submarine is squeezed from all sides by this pressure. A deep-sea diving submarine must be able to withstand the terrific force of water at great depths. Therefore, the air pressure within it must be equal to the hydrostatic pressure surrounding it.

PRINCIPLES OF HYDROSTATIC PRESSURE

In chapter 9 you found out that all fluids exert pressure in all directions. That's simple enough. How great is the pressure? Try a little experiment. Place a pile of blocks in front of you on the table. Stick the tip of your finger under the first block from the top. Not much pressure on your finger, is there? Stick it between the third and fourth blocks. The pressure on your finger has increased. Now slide your finger under the bottom block in the pile. There you will find the pressure is greatest. The pressure increases as you go lower in the pile. You might say that pressure increases with depth. The same

is true in liquids. The deeper you go, the greater the pressure becomes. However, depth isn't the whole story.

Suppose the blocks in the preceding paragraph were made of lead. The pressure at any level in the pile would be considerably greater. Or suppose they were blocks of balsa wood-then the pressure at each level wouldn't be as great. Pressure, then, depends not only on the depth, but also on the weight of the material. Since you are dealing with pressure—force per unit of area, you will also be dealing with weight per unit of volume-or density.

When you talk about the density of a substance, you are talking about its weight per cubic foot or per cubic inch. For example, the density of water is 62.5 pounds per cubic foot; the density of lead is 710 pounds per cubic foot. However, to say that lead is heavier than water isn't a true statement. For instance, a 22-caliber bullet is the same density as a pail of water, but the pail of water is much heavier. It is true, however, that a cubic foot of lead is much heavier than a cubic foot of water.

Pressure depends on two principles-depth and density. You can easily find the pressure at any depth in any liquid by using the following formula:

$$P = H \times D$$

in which

P = pressure, in lb per sq in. or lb per sq ft

H = depth of the point, measured in feet or inches

and

D = density in lb per cu in. or lb per cu ft

Note: If you use inches in your computation, you must use them throughout; if you use feet, you must use them throughout.

What is the pressure on 1 square foot of the surface of a submarine if the submarine is 200 feet below the surface? Using the formula:

$$P = H \times D$$

$$P = 200 \times 62.5 = 12,500 \text{ lb per sq ft}$$

Every square foot of the sub's surface that is at that depth has a force of more than 6 tons pushing in on it. If the height of the hull is 20 feet and the area in question is between the sub's top and bottom, you can see that the pressure on the hull will be at least $(200 - 10) \times 62.5 = 11,875$ pounds per square foot. The greatest pressure will be $(200 + 10) \times 62.5 = 13,125$ pounds per square foot. Obviously, the hull has to be very strong to withstand such pressures.

USES OF HYDROSTATIC PRESSURE

Various shipboard operations depend on the use of hydrostatic pressure. For example, in handling depth charges, torpedoes, mines, and some types of aerial bombs, you'll be dealing with devices that operate by hydrostatic pressure. In addition, you'll deal with hydrostatic pressure in operations involving divers.

Firing Depth Charges

Hiding below the surface exposes the submarine to great fluid pressure. However, it also gives the sub a great advantage because it is hard to hit and, therefore, hard to kill. A depth charge must explode within 30 to 50 feet of a submarine to cause damage. That means the depth charge must not go off until it has had time to sink to approximately the same level as the sub. Therefore, you use a firing mechanism that is set off by the pressure at the estimated depth of the submarine.

Figure 10-1 shows a depth charge and its interior components. A depth charge is a sheet-metal container filled with a high explosive and a firing device. A tube passes through its center from end to end. Fitted in one end of this tube is the booster, a load of granular TNT that sets off the main charge. It is also fitted with a safety fork and an inlet valve cover. Upon launching, the safety fork is knocked off, and the valve cover is removed to allow water to enter.

When the depth charge gets about 12 to 15 feet below the surface, the water pressure is sufficient to extend a bellows in the booster extender. The bellows

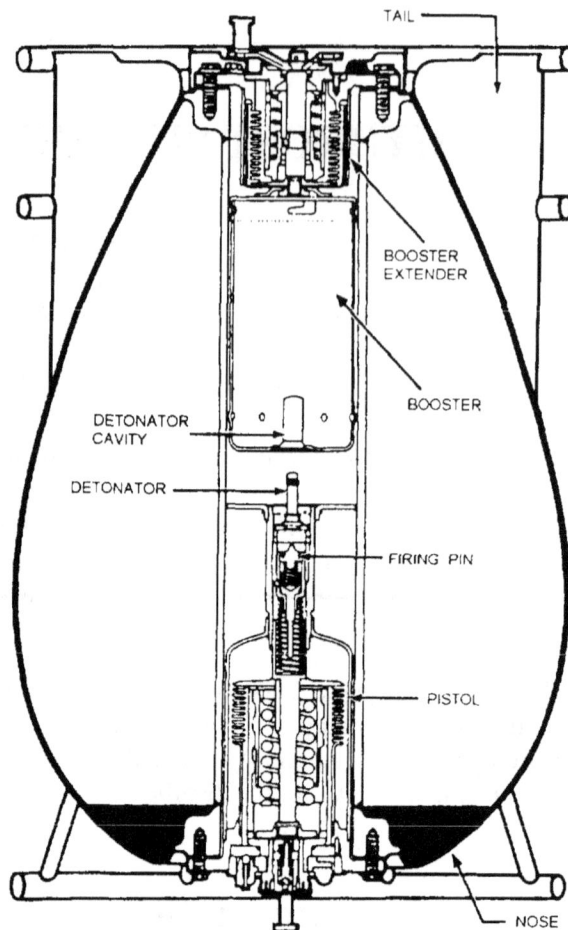

Figure 10-1.-A depth charge.

trips a release mechanism, and a spring pushes the booster up against the centering flange. Notice that the detonator fits into a pocket in the booster. Unless the detonator is in this pocket, it cannot set off the booster charge.

Nothing further happens until the detonator fires. As you can see, the detonator fits into the end of the pistol, with the firing pin aimed at the detonator base. The pistol also contains a bellows into which the water rushes as the charge goes down. As the pressure increases, the bellows begins to expand against the depth spring. You can adjust this spring so that the bellows will have to exert a predetermined force to compress it.

Figure 10-2 shows you the depth-setting dials of one type of depth charge. Since the pressure on the bellows depends directly on the depth, you can select any depth on the dial at which you wish the charge to go off. When the pressure in the bellows becomes sufficiently great, it releases the firing spring, which drives the firing pin

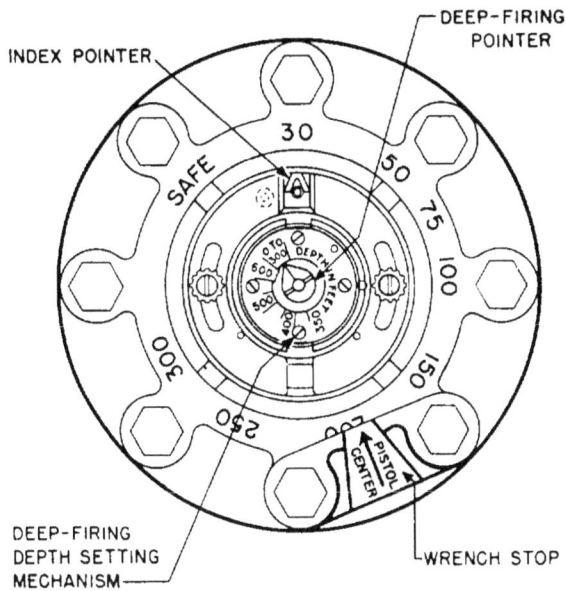

Figure 10-2.-Depth-setting dial.

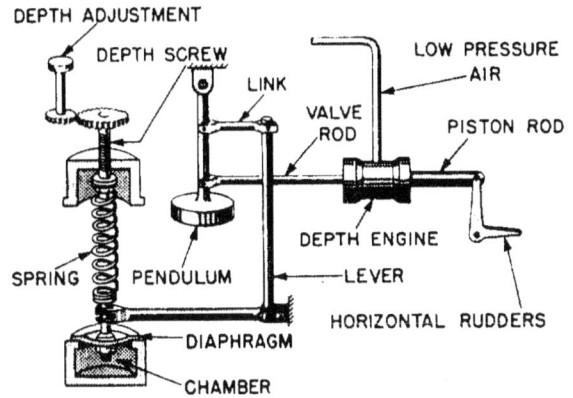

Figure 10-3.-Inside a torpedo.

into the detonator. The booster, already in position, then fires and, in turn, sets off the entire load of TNT.

These two bellows—operated by hydrostatic pressure—serve two purposes. First, they permit the depth charge to fire at the proper depth; second, they make the charge safe to handle and carry. If you should accidentally knock the safety fork and the valve inlet cover off on deck, nothing would happen. Even if the detonator should go off while you were handling the charge, the main charge would not fire unless the booster was in the extended position.

Guiding Torpedoes

To keep a torpedo on course toward its target is a job. Maintaining the proper compass course with a gyroscope is only part of the problem. The torpedo must travel at the proper depth so that it will neither pass under the target ship nor hop out of the water on the way.

As figure 10-3 shows, the torpedo contains an air-filled chamber sealed with a thin, flexible metal plate, or diaphragm. This diaphragm can bend upward or downward against the spring. You determine the spring tension by setting the depth-adjusting knob.

Suppose the torpedo starts to dive below the selected depth. The water, which enters the torpedo and surrounds the chamber, exerts an increased pressure on the diaphragm and causes it to bend down. If you follow the lever system, you can see that the pendulum will push forward. Notice that a valve rod connects the

pendulum to the piston of the depth engine. As the piston moves to the left, low-pressure air from the torpedo's air supply enters the depth engine to the right of the piston and pushes it to the left. You must use a depth engine because the diaphragm is not strong enough to move the rudders.

The piston of the depth engine connects to the horizontal rudders as shown. When the piston moves to the left, the rudder turns upward and the torpedo begins to rise to the proper depth. If the nose goes up, the pendulum swings backward and keeps the rudder from elevating the torpedo too rapidly. As long as the torpedo runs at the selected depth, the pressure on the chamber remains constant and the rudders do not change from their horizontal position.

Diving

Navy divers have a practical, first-hand knowledge of hydrostatic pressure. Think what happens to divers who go down 100 feet to work on a salvage job. The pressure on them at that depth is 8,524 pounds per square foot! Something must be done about that, or they would be flatter than a pancake.

To counterbalance this external pressure, a diver wears a rubber suit. A shipboard compressor then pumps pressurized air into the suit, which inflates it and provides oxygen to the diver's body as well. The oxygen enters the diver's lungs and bloodstream, which carries it to every part of the body. In that way the diver's internal pressure is equal to the hydrostatic pressure.

As the diver goes deeper, the air pressure increases to meet that of the water. In coming up, the pressure on the air is gradually reduced. If brought up too rapidly, the diver gets the "bends." That is, the air that was dissolved in the blood begins to come out of solution and form bubbles in the veins. Any sudden release in the pressure on a fluid results in the freeing of some gases that are dissolved in the fluid. You have seen this happen when you suddenly relieve the pressure on a bottle of pop by removing the cap. The careful matching of hydrostatic pressure on the diver by air pressure in the diving suit is essential if diving is to be done at all.

Determining Ship's Speed

Did you ever wonder how the skipper knows the speed the ship is making through water? The skipper can get this information by using several instruments-the patent log, the engine revolution counter, and the pitometer (pit) log. The "pit log" operates, in part, by hydrostatic pressure. It really shows the difference between hydrostatic pressure and the pressure of the water flowing past the ship-but this difference can be used to find ship's speed.

Figure 10-4 shows a schematic drawing of a pitometer log. It consists of a double-wall tube that sticks out forward of the ship's hull into water that is not disturbed by the ship's motion. In the tip of the tube is an opening (A). When the ship is moving, two forces or pressures are acting on this opening: (1) the hydrostatic pressure caused by the depth of the water above the opening and (2) a pressure caused by the push of the ship through the water. The total pressure from these two forces transmits through the central tube (shown in white on the figure) to the left-hand arm of a manometer.

In the side of the tube is a second opening (B) that does not face the direction in which the ship is moving. Opening B passes through the outer wall of the double-wall tube, but not through the inner wall. The only pressure affecting opening B is the hydrostatic

pressure. This pressure transmits through the outer tube (shaded in the drawing) to the right-hand arm of the manometer.

When the ship is dead in the water, the pressure through both openings A and B is the same, and the mercury in each arm of the manometer stands at the same level. However, as soon as the ship begins to move, additional pressure develops at opening A, and the mercury pushes down in the left-hand arm and up into the right-hand arm of the tube. The faster the ship goes, the greater this additional pressure becomes, and the greater the difference will be between the levels of the mercury in the two arms of the manometer. You can read the speed of the ship directly from the calibrated scale on the manometer.

Since air is also a fluid, the airspeed of an aircraft can be found by a similar device. You have probably seen the thin tube sticking out from the nose or the leading edge of a wing of the plane. Flyers call this tube a pitot tube. Its basic principle is the same as that of the pitometer log.

HYDRAULIC PRESSURE

Perhaps your earliest contact with hydraulic pressure was when you got your first haircut. The hairdresser put a board across the arms of the chair, sat you on it, and began to pump the chair up to a convenient level. As you grew older, you probably discovered that the gas station attendant could put a car on the greasing rack and-by some mysterious arrangement-jack it head high. The attendant may have told you that oil under pressure below the piston was doing the job.

Come to think about it, you've probably known something about hydraulics for a long time. Automobiles and airplanes use hydraulic brakes. As a sailor, you'll have to operate many hydraulic machines. You'll want to understand the basic principles on which they work.

Primitive man used simple machines such as the lever, the inclined plane, the pulley, the wedge, and the wheel and axle. It was considerably later before someone discovered that you could use liquids and gases to exert forces at a distance. Then, a vast number of new machines appeared. A machine that transmits forces by a liquid is a hydraulic machine. A variation of the hydraulic machine is the type that operates with a compressed gas. This type is known as the pneumatic machine. This chapter deals only with basic hydraulic machines.

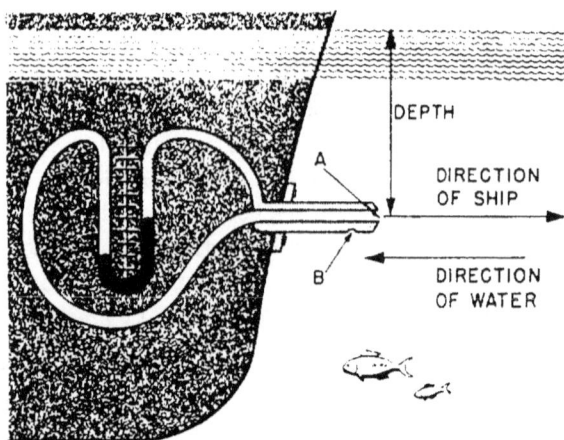

figure 10-4.-A pitometer log.

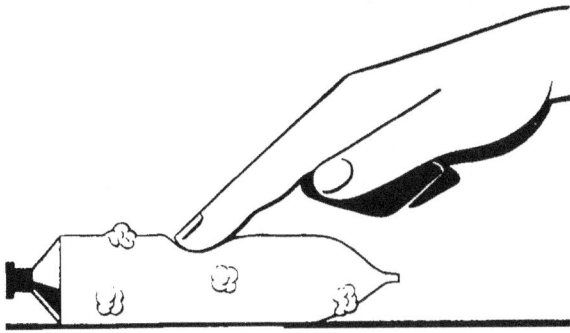

Figure 10-5.-Pressure to a fluid transmits in all directions.

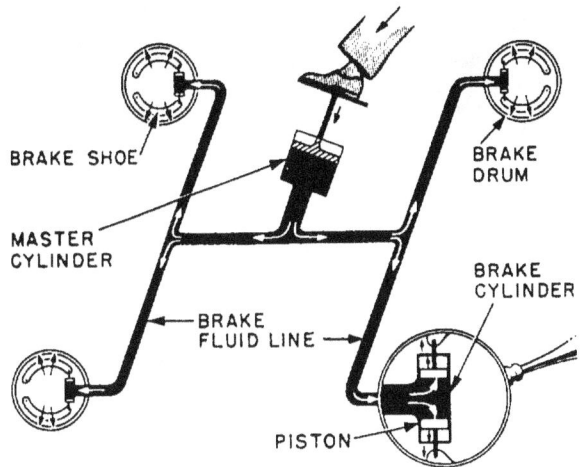

Figure 10-6.-Hydraulic brakes.

PRINCIPLES OF HYDRAULIC PRESSURE

A Frenchman named Pascal discovered that a pressure applied to any part of a confined fluid transmits to every other part with no loss. The pressure acts with equal force on all equal areas of the confining walls and perpendicular to the walls.

Remember when you are talking about the hydraulic machine, you are talking about the way a liquid acts in a closed system of pipes and cylinders. The action of a liquid under such conditions is somewhat different from its behavior in open containers or in lakes, rivers, or oceans. You also should keep in mind that you cannot compress most liquids into a smaller space. Liquids don't "give" the way air does when you apply pressure, nor do liquids expand when you remove pressure.

Punch a hole in a tube of toothpaste. If you push down at any point on the tube, the toothpaste comes out of the hole. Your force has transmitted from one place to another through the toothpaste, which is a thick, liquid fluid. Figure 10-5 shows what would happen if you punched four holes in the tube. If you were to press on the tube at one point, the toothpaste would come out of all four holes. You have illustrated a basic principle of hydraulic machines. That is, a force applied on a liquid transmits equally in every direction to all parts of the container.

We use this principle in the operation of four-wheel hydraulic automobile brakes. Figure 10-6 is a simplified drawing of this brake system. You push down on the brake pedal and force the piston in the master cylinder against the fluid in that cylinder. This push sets up a pressure on the fluid as your finger did on the toothpaste in the tube. The pressure on the fluid in the master cylinder transmits through the lines to the brake cylinders in each wheel. This fluid under pressure

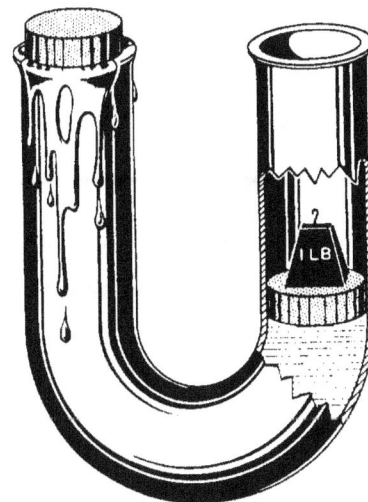

Figure 10-7.-Liquid transmits force.

pushes against the pistons in each of the brake cylinders and forces the brake shoes out against the drums.

MECHANICAL ADVANTAGES OF HYDRAULIC PRESSURE

Another aspect to understand about hydraulic machines is the relationship between the force you apply and the result you get. Figure 10-7 will help you understand this principle. The U-shaped tube has a cross-sectional area of 1 square inch. In each arm is a piston that fits snugly, but can move up and down. If you place a 1-pound weight on one piston, the other one will push out the top of its arm immediately. If you place a

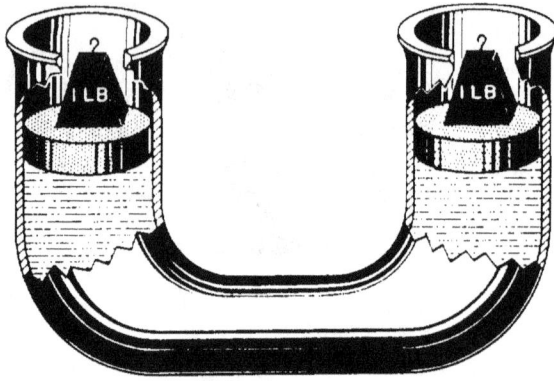

Figure 10-8.-Equal pressure applied at each end of a tube containing a liquid.

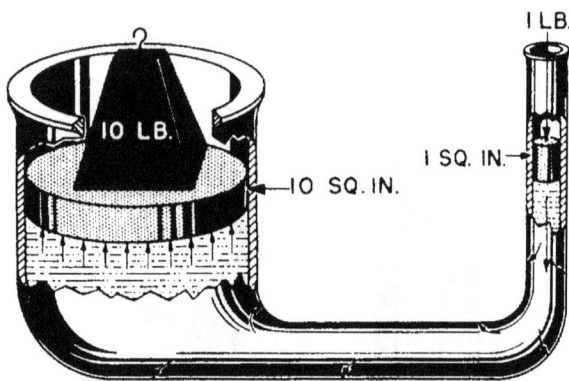

Figure 10-9.-A mechanical advantage of 10.

1-pound weight on each piston, however, each one will remain in its original position, as shown in figure 10-8.

Thus, you see that a pressure of 1 pound per square inch applied downward on the right-hand piston exerts a pressure of 1 pound per square inch upward against the left-hand one. Not only does the force transmit through the liquid around the curve, it transmits equally on each unit area of the container. It makes no difference how long the connecting tube is or how many turns it makes. It is important that the entire system be full of liquid. Hydraulic systems will fail to operate properly if air is present in the lines or cylinders.

Now look at figure 10-9. The piston on the right has an area of 1 square inch, but the piston on the left has an area of 10 square inches. If you push down on the smaller piston with a force of 1 pound, the liquid will transmit this pressure to every square inch of surface in the system. Since the left-hand piston has an area of 10 square inches, each square inch has a force of 1 pound

transmitted to it. The total effect is a push on the larger piston with a total force of 10 pounds. Set a 10-pound weight on the larger piston and it will support the 1-pound force of the smaller piston. You then have a 1-pound push resulting in a 10-pound force. That's a mechanical advantage of 10. This mechanical advantage is why hydraulic machines are important.

Here's a formula that will help you to figure the forces that act in a hydraulic machine:

$$\frac{F_1}{F_2} = \frac{A_1}{A_2}$$

In that,

F_1 = force, in pounds, applied to the small piston;

F_2 = force, in pounds, applied to the large piston;

A_1 = area of the small piston, in square inches; and

A_2 = area of the large piston, in square inches.

Let's apply the formula to the hydraulic press shown in figure 10-10. The large piston has an area of 90 square inches, and the smaller one has an area of 2 square inches. The handle exerts a total force of 15 pounds on the small piston. With what total force could you raise the large piston?

Write down the formula

$$\frac{F_1}{F_1} = \frac{A_1}{A_2}.$$

Substitute the known values

$$\frac{15}{F_2} = \frac{2}{90}$$

and

$$F_2 = \frac{90 \times 15}{2} = 675 \text{ pounds.}$$

USES OF HYDRAULIC PRESSURE

You know from your experience with levers that you can't get something for nothing. Applying this knowledge to the simple system in figure 10-9, you know that you can't get a 10-pound force from a

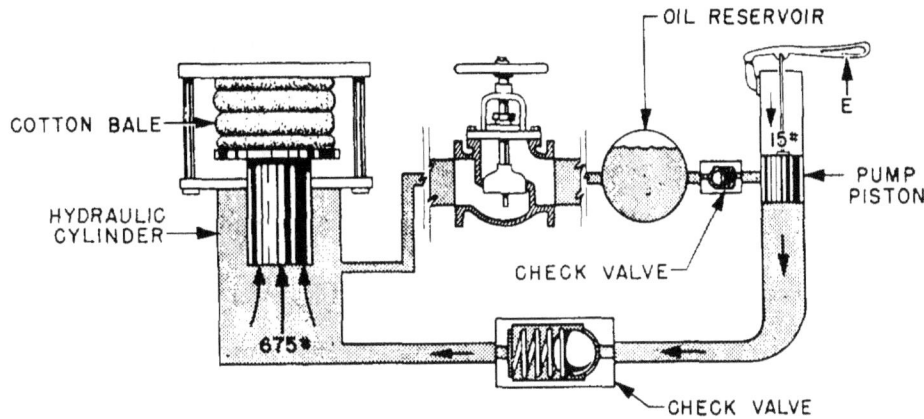

Figure 10-10.-Hydraulic press.

1-pound effort without sacrificing distance. You must apply the 1-pound effort through a much greater distance than the 10-pound force will move. To raise the 10-pound weight a distance of 1 foot, you must apply the 1-pound effort through what distance? Remember, if you neglect friction, the work done on any machine equals the work done by that machine. Use the work formula to find how far the smaller piston will have to move.

Work input = Work output

$$F_1 \times D_1 = F_2 \times D_2$$

By substituting

$$1 \times D_1 = 10 \times 1$$

you find that

$$D_1 = 10 \text{ feet}$$

The smaller piston will have to move a distance of 10 feet to raise the 10-pound load 1 foot. It looks then as though the smaller cylinder would have to be at least 10 feet long—and that wouldn't be practical. In addition, it isn't necessary if you put a valve in the system.

The hydraulic press in figure 10-10 contains a valve. As the small piston moves down, it forces the fluid past check valve A into the large cylinder. As soon as the small piston moves upward, it removes the pressure to the right of check valve A. The pressure of the fluid on the check valve spring below the large piston helps force

that valve shut. The liquid that has passed through the valve opening on the down stroke of the small piston is trapped in the large cylinder.

The small piston rises on the upstroke until its bottom passes the opening to the fluid reservoir. More fluid is sucked past check valve B and into the small cylinder. The next downstroke forces this new charge of fluid out of the small cylinder past the check valve into the large cylinder. This process repeats stroke by stroke until enough fluid has been forced into the large cylinder to raise the large piston the required distance of 1 foot. The force has been applied through a distance of 10 feet on the pump handle. However, it was done through a series of relatively short strokes, the total of the strokes being equal to 10 feet.

Maybe you're beginning to wonder how the large piston gets back down after the process is finished. The fluid can't run back past check valve B-that's obvious, Therefore, you lower the piston by letting the oil flow back into the reservoir through a return line. Notice that a simple globe valve is in this line. When the globe valve opens, the fluid flows back into the reservoir. Of course, this valve is shut while the pump is in operation.

Aiding the Helmsman

You've probably seen the helmsman swing a ship weighing thousands of tons almost as easily as you turn your car. No, helmsmen are not superhuman. They control the ship with machines. Many of these machines are hydraulic.

There are several types of hydraulic and electro-hydraulic steering mechanisms. The simplified diagram

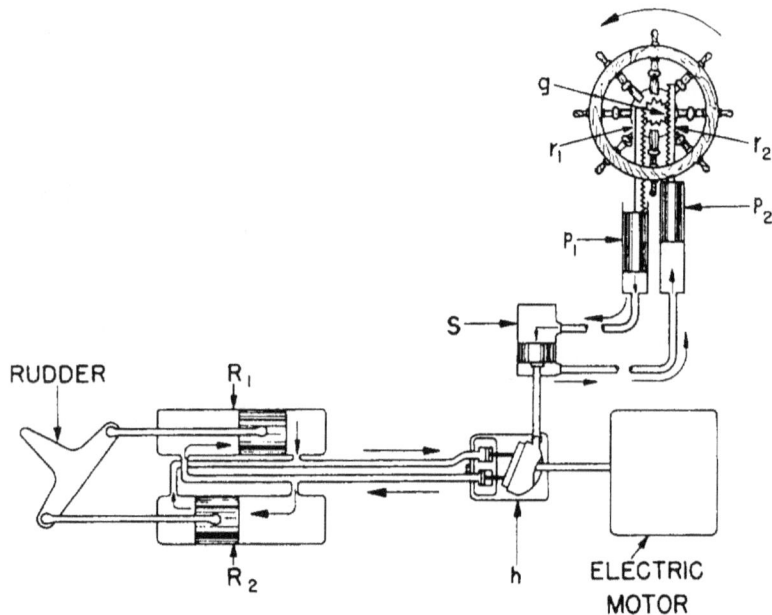

Figure 10-11.-Electrohydraulic steering mechanism.

in figure 10-11 will help you to understand the general principles of their operation. As the hand steering wheel turns in a counterclockwise direction, its motion turns the pinion gear (g). This causes the left-hand rack (r_1) to move downward and the right-hand rack (r_2) to move upward. Notice that each rack attaches to a piston (p_1 or p_2). The downward motion of rack r_1 moves piston p_1 downward in its cylinder and pushes the oil out of that cylinder through the line. At the same time, piston p_2 moves upward and pulls oil from the right-hand line into the right-hand cylinder.

If you follow these two lines, you see that they enter a hydraulic cylinder (S). One line enters above and one below the single piston in that cylinder. This piston and the attached plunger are pushed down toward the hydraulic pump (h) in the direction of the oil flow shown in the diagram. So far in this operation, hand power has been used to develop enough oil pressure to move the control plunger attached to the hydraulic pump. At this point, an electric motor takes over and drives the pump (h).

Oil is pumped under pressure to the two big steering rams (R_1 and R_2). You can see that the pistons in these rams connect directly to the rudder crosshead that controls the position of the rudder. With the pump operating in the direction shown, the ship's rudder is thrown to the left, and the bow will swing to port. This operation shows how a small force applied on the steering wheel sets in motion a series of operations that result in a force of thousands of pounds.

Getting Planes on Deck

The swift, smooth power required to get airplanes from the hanger deck to the flight deck of a carrier is provided by a hydraulic lift. Figure 10-12 shows how this lifting is done. An electric motor drives a variable-speed gear pump. Oil enters the pump from the reservoir and is forced through the lines to four hydraulic rams. The pistons of the rams raise the elevator platform. The oil under pressure exerts its force on each square inch of surface area of the four pistons. Since the pistons are large, a large total lifting force results. Either reversing the pump or opening valve 1 and closing valve 2 lowers the elevator. The weight of the elevator then forces the oil out of the cylinders and back into the reservoir.

Operating Submarines

Another application of hydraulics is the operation of submarines. Inside a submarine, between the outer skin and the pressure hull, are several tanks of various design and purpose. These tanks control the total weight of the ship, allowing it to submerge or surface. They also control the trim or balance, fore and aft, of the submarine. The main ballast tanks have the primary function of either destroying or restoring positive buoyancy to the submarine. Allowing air to escape through hydraulically operated vents at the top of the tanks lets seawater enter through the flood ports at the bottom to replace the air. For the sub to regain positive buoyancy, the tanks are "blown" free of seawater with

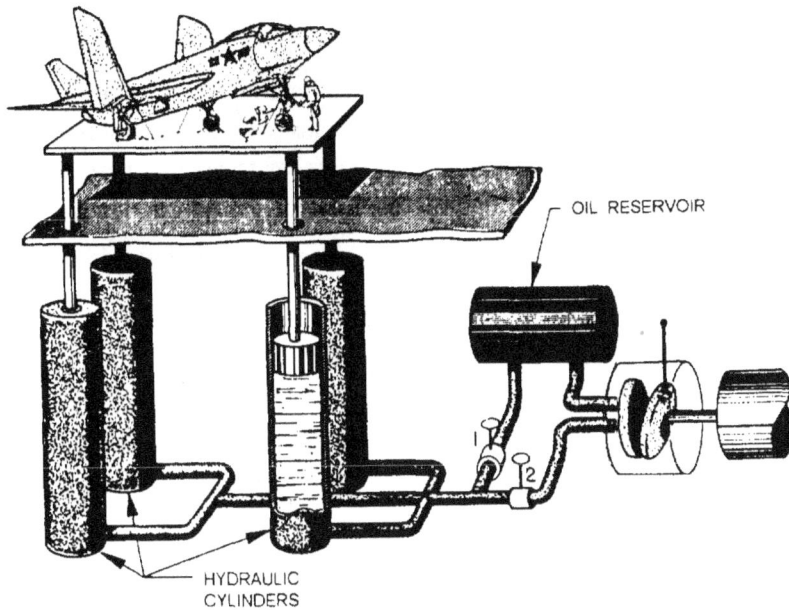

Figure 10-12.-Hydraulic lift.

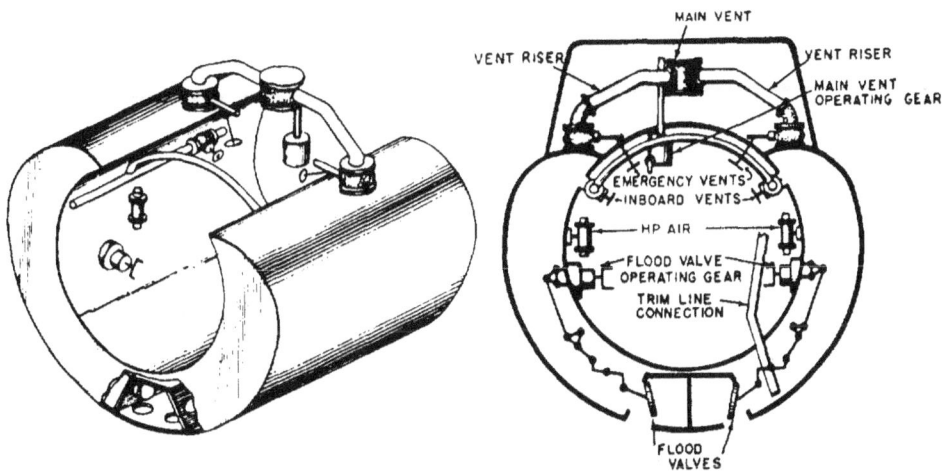

Figure 10-13.-Submarine special ballast tank (safety tank).

compressed air. Sufficient air is left trapped in the tanks to prevent the seawater from reentering.

We use other tanks, such as variable ballast tanks and special ballast tanks (for example, the negative tank, safety tank, and bow buoyancy tank), either for controlling trim or stability or for emergency weight-compensating purposes. The variable ballast tanks have no direct connection to the sea. Therefore, we must pump water into or out of them. The negative tank and the safety tank can open to the sea through large flood valves. These valves, as well as the vent valves for the main ballast tanks and those for the safety and negative tanks, are all hydraulically operated.

The vents and flood valves are outside the pressure hull, so some means of remote control is needed to open and close them from within the submarine. We use hydraulic pumps, lines, and rams for this purpose. Oil pumped through tubing running through the pressure hull actuates the valve's operating mechanisms by exerting pressure on and moving a piston in a hydraulic cylinder. Operating the valves by a hydraulic system from a control room is easier and simpler than doing so by a mechanical system of gears, shafts, and levers. The hydraulic lines can be readily led around corners and obstructions, and a minimum of moving parts is required.

Figure 10-13 is a schematic sketch of the safety tank-one of the special ballast tanks in a submarine. The main vent and the flood valves of this tank operate

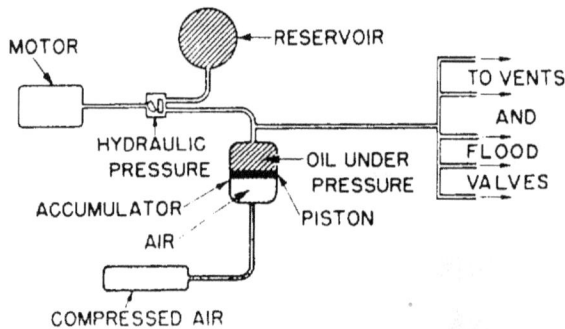

Figure 10-14.-Controlling fluid pressure.

hydraulically by remote control, although in an emergency they may operate manually.

Hydraulics are used in many other ways aboard submarines. They are used to raise and lower the periscope. The submarines are steered and the bow and stern planes are controlled by hydraulic systems. The windlass and capstan system, used in mooring the submarine, is hydraulically operated. You will find many more applications of hydraulics aboard the submarine.

Controlling Fluid Pressure

In some hydraulic systems, oil is kept under pressure in a container known as an accumulator. As shown in figure 11-14, the accumulator is a large cylinder; oil is pumped into it from the top. A free piston divides the cylinder into two parts. Compressed air is forced into the cylinder below the piston at a pressure of 600 psi. Oil is then forced into it on top of the piston. As the pressure above it increases, the piston is forced down, squeezing the air into a smaller space. Air is elastic; you can compress it under pressure, and it will expand as soon as the pressure is reduced. When oil pressure is reduced, large quantities of oil under working pressure are instantly available to operate hydraulic rams or motors any place on the submarine.

SUMMARY

The Navy uses many devices whose operation depends on the hydrostatic principle. You should remember three points about the operation of these devices:

Pressure in a liquid is exerted equally in all directions.

Hydrostatic pressure refers to pressure at any depth in a liquid that is not flowing.

Pressure depends upon both depth and density.

The formula for finding pressure is

$P = H \times D$

The working principle of all hydraulic mechanisms is simple enough. Whenever you find an application that seems hard to understand, keep these points in mind:

Hydraulics is the term applied to the behavior of enclosed liquids. Machines that operate liquids under pressure are called hydraulic machines.

Liquids are incompressible. They cannot be squeezed into spaces smaller than they originally occupied.

A force applied on any area of a confined liquid transmits equally to every part of that liquid.

In hydraulic cylinders, the relation between the force exerted by the large piston to the force applied on the smaller piston is the same as the relationship between the area of the larger piston and the area of the smaller piston.

Some of the advantages of hydraulic machines are:

We use tubing to transmit forces, and tubing can readily transmit forces around corners.

Tubing requires little space.

Few moving parts are required.

CHAPTER 11

MACHINE ELEMENTS AND BASIC MECHANISMS

CHAPTER LEARNING OBJECTIVES

Upon completion of this chapter, you should be able to do the following:

- *Describe the machine elements used in naval machinery and equipment.*

- *Identify the basic machines used in naval machiney and equipment.*

- *Explain the use of clutches.*

Any machine, however simple, consists of one or more basic machine elements or mechanisms. In this chapter we will take a look at some of the more familiar elements and mechanisms used in naval machinery and equipment.

BEARINGS

Friction is the resistance of force between two surfaces. In chapter 7 we saw that two objects rubbing against each other produce friction. If the surfaces are smooth, they produce little friction; if either or both are rough, they produce more friction. To start rolling a loaded hand truck across the deck, you would have to give it a hard tug to overcome the resistance of static friction. To start sliding the same load across the deck, you would have to give it an even harder push. That is because rolling friction is always less than sliding friction. We take advantage of this fact by using rollers or bearings in machines to reduce friction. We use lubricants on bearing surfaces to reduce the friction even further.

A bearing is a support and guide that carries a moving part (or parts) of a machine. It maintains the proper relationship between the moving part or parts and the stationary part. It usually permits only one form of motion, such as rotation. There are two basic types of bearings: sliding (plain bearings), also called friction or

guide bearings, and antifrictional (roller and ball bearings).

SLIDING BEARINGS

In sliding (plain) bearings, a film of lubricant separates the moving part from the stationary part. Three types of sliding bearings are commonly used: reciprocal motion bearings, journal bearings, and thrust bearings.

Reciprocal Motion Bearings

Reciprocal motion bearings provide a bearing surface on which an object slides back and forth. They are found on steam reciprocating pumps, in which connecting rods slide on bearing surfaces near their connections to the pistons. We use similar bearings on the connecting rods of large internal-combustion engines and in many mechanisms operated by cams.

Journal Bearings

Journal bearings guide and support revolving shafts. The shaft revolves in a housing fitted with a liner. The inside of the liner, on which the shaft bears, is made of babbitt metal or a similar soft alloy (antifriction metal) to reduce friction. The soft metal is backed by a bronze or copper layer and has a steel back for strength. Sometimes the bearing is made in two halves and is

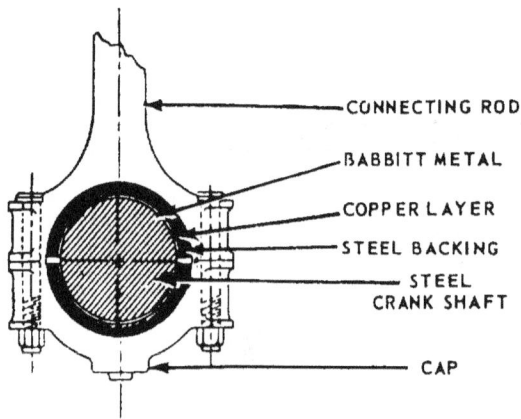

Figure 11-1.-Babbitt-lined bearing in which steel shaft revolves.

Figure 11-3.-Diagrammatic arrangement of a Kingsbury thrust bearing, showing oil film.

clamped or screwed around the shaft (fig. 11-1). We also call it a laminated sleeve bearing.

Under favorable conditions the friction in journal bearings is remarkably small. However, when the rubbing speed of a journal bearing is very low or extremely high, the friction loss may become excessive. A good example is the railroad car. Railroad cars are now being fitted with roller bearings to eliminate the "hot box" troubles associated with journal bearings.

Heavy-duty bearings have oil circulated around and through them. Some have an additional cooling system that circulates water around the bearing. Although revolving the steel shaft against babbitt metal produces less friction (and less heat and wear) than steel against

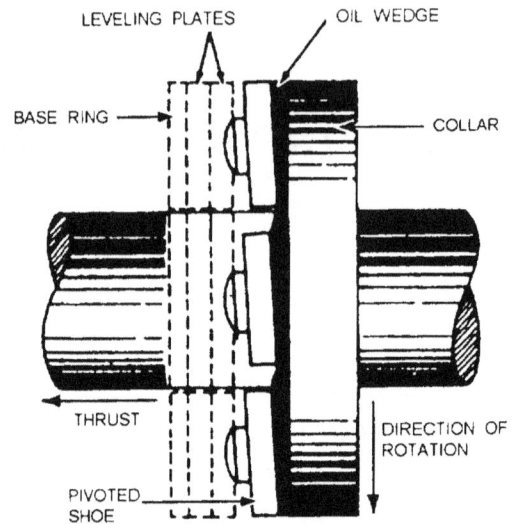

steel, keeping the parts cool is still a problem. The same care and lubrication needed to prevent a burned out bearing on your car is needed on all Navy equipment, only more so. Many lives depend on the continued operation of Navy equipment.

Thrust Bearings

Thrust bearings are used on rotating shafts, such as those supporting bevel gears, worm gears, propellers, and fans. They resist axial thrust or force and limit axial

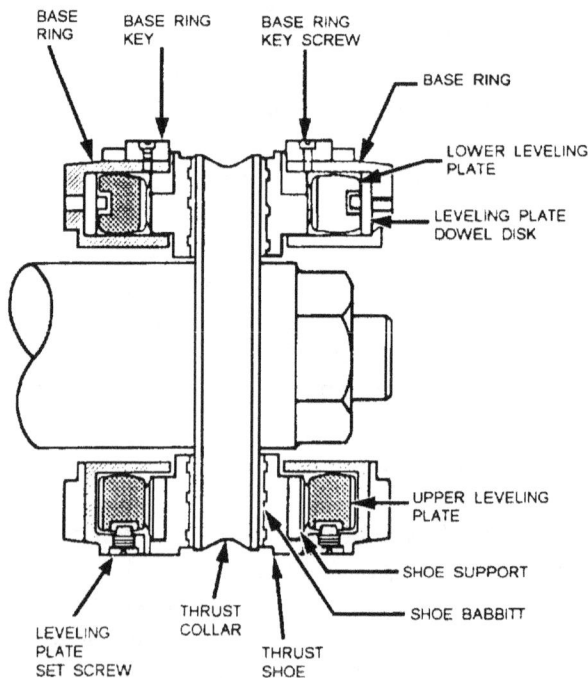

Figure 11-2.-Kingsbury pivoted-shoe thrust bearing.

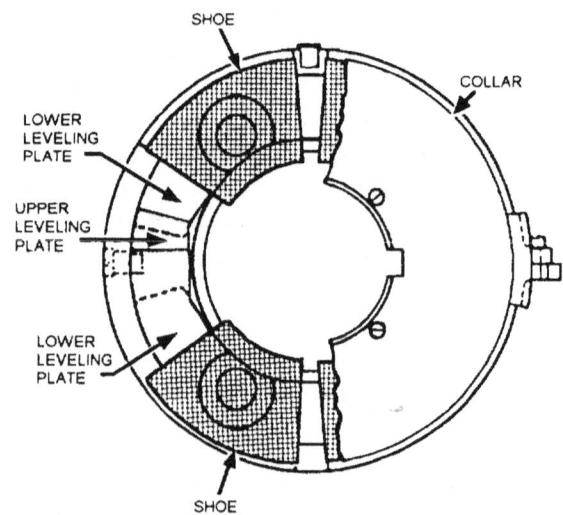

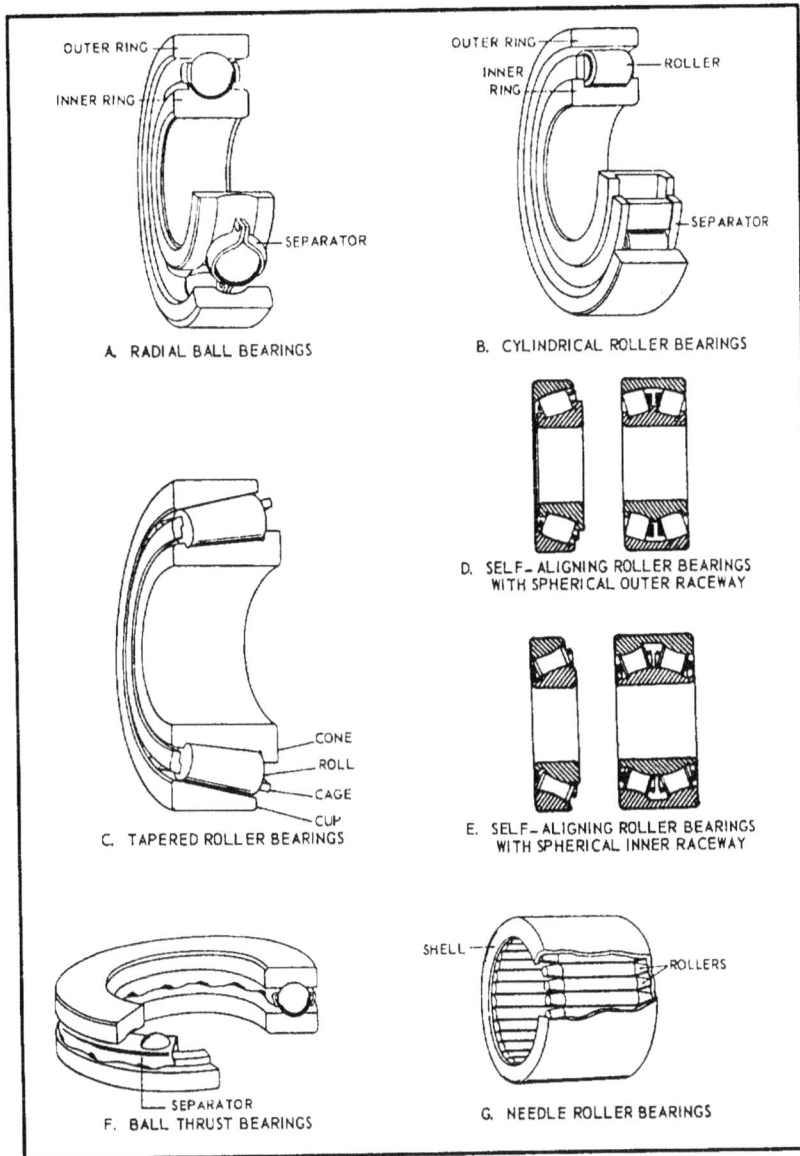

Figure 11-4.-The seven basic types of antifrictional hearings.

movement. They are used chiefly on heavy machinery, such as Kingsbury thrust bearings used in heavy marine-propelling machinery (figs. 11-2 and 11-3). The base of the housing holds an oil bath, and the rotation of the shaft continually distributes the oil. The bearing consists of a thrust collar on the propeller shaft and two or more stationary thrust shoes on either side of the collar. Thrust is transmitted from the collar through the shoes to the gear housing and the ship's structure to which the gear housing is bolted.

ANTIFRICTIONAL OR ROLLER AND BALL BEARINGS

You have had first-hand acquaintance with ball bearings since you were a child. They are what made your roller skates or bicycle wheels spin freely. If any of the little steel balls came out and were lost, your roller skates screeched and groaned.

Antifrictional balls or rollers are made of hard, highly polished steel. Typical bearings consist of two hardened steel rings (called races), the hardened steel balls or rollers, and a separator. The motion occurs between the race surfaces and the rolling elements. There are seven basic types of antifrictional bearings (fig. 11-4):

1. Radial ball bearings

2. Cylindrical roller bearings

3. Tapered roller bearings

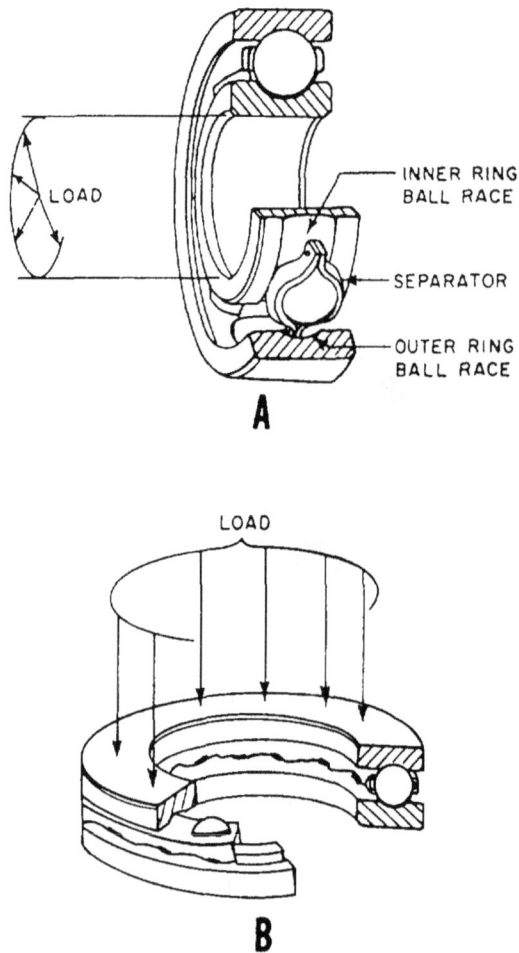

Figure 11-5.-Ball bearings. A. Radial type; B. Thrust type.

4. Self-aligning roller bearings with a spherical outer raceway

5. Self-aligning roller bearings with a spherical inner raceway

6. Ball thrust bearings

7. Needle roller bearings

Roller bearing assemblies are usually easy to disassemble for inspection, cleaning, and replacement of parts. Ball bearings are assembled by the manufacturer and are installed, or replaced, as a unit. Sometimes maintenance publications refer to roller and ball bearings as either trust or radial bearings. The difference between the two depends on the angle of intersection between the direction of the load and the plane of rotation of the bearing.

Figure 11-5, A, shows a radial ball bearing assembly. The load shown is pressing outward along the radius of the shaft. Now suppose a strong thrust were to be exerted on the right end of the shaft in an effort to

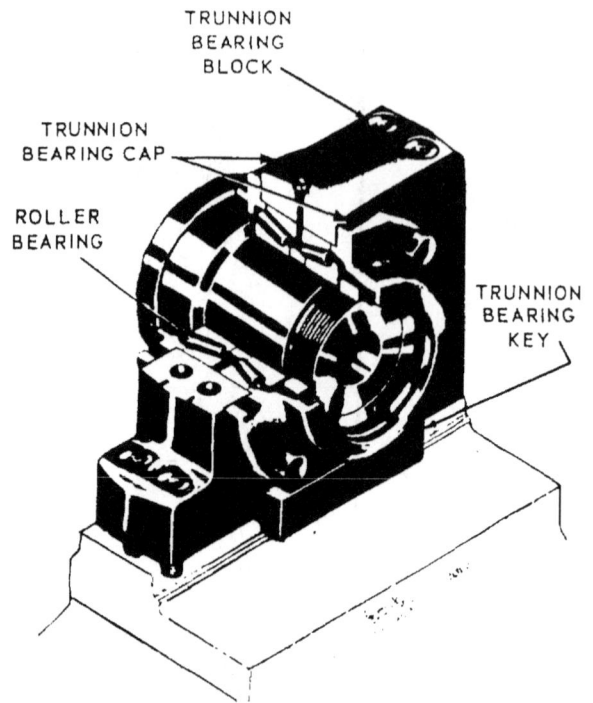

Figure 11-6.-Radial-thrust roller bearing.

move it to the left. You would find that the radial bearing is not designed to support this axial thrust. Even putting a shoulder between the load and the inner race wouldn't support it; instead, the bearings would pop out of their races.

Supporting a thrust on the right end of the shaft would require the thrust bearing arrangement of the braces shown in figure 11-5, B. A shoulder under the lower race and another between the load and the upper race would handle any axial load up to the design limit of the bearing.

Sometimes bearings are designed to support both thrust and radial loads. This explains the use of the term "radial thrust" bearings. The tapered roller bearing in figure 11-6 is an example of a radial-thrust roller bearing.

Antifriction bearings require smaller housings than other bearings of the same load capacity and can operate at higher speeds.

SPRINGS

Springs are elastic bodies (generally metal) that can be twisted, pulled, or stretched by some force. They can return to their original shape when the force is released. All springs used in naval machinery are made of metal—usually steel—though some are made of phosphor bronze, brass, or other alloys. A part that is subject to constant spring thrust or pressure is said to be

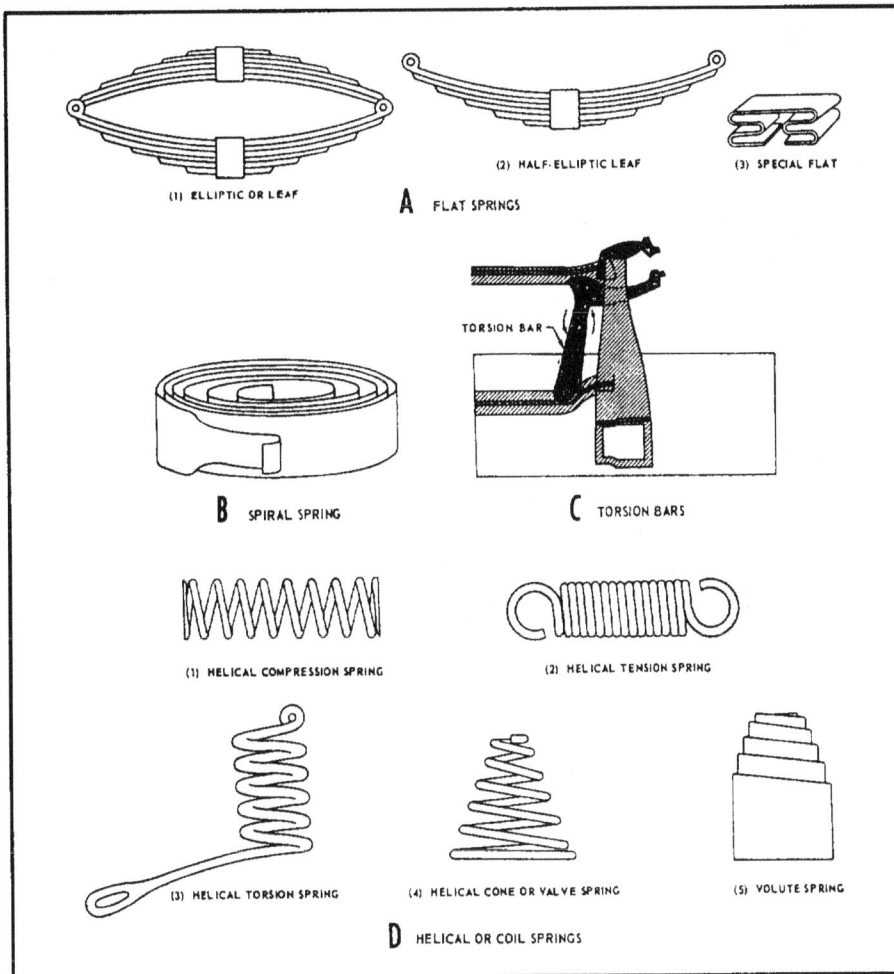

Figure 11-7.-Types of springs.

spring-loaded. (Some components that appear to be spring-loaded are actually under hydraulic or pneumatic pressure or are moved by weights.)

FUNCTIONS OF SPRINGS

Springs are used for many purposes, and one spring may serve more than one purpose. Listed below are some of the more common of these functional purposes. As you read them, try to think of at least one familiar application of each.

1. To store energy for part of a functioning cycle.

2. To force a component to bear against, to maintain contact with, to engage, to disengage, or to remain clear of some other component.

3. To counterbalance a weight or thrust (gravitational, hydraulic, etc.). Such springs are usually called equilibrator springs.

4. To maintain electrical continuity.

5. To return a component to its original position after displacement.

6. To reduce shock or impact by gradually checking the motion of a moving weight.

7. To permit some freedom of movement between aligned components without disengaging them. These are sometimes called take-up springs.

TYPES OF SPRINGS

As you read different books, you will find that authors do not agree on the classification of types of springs. The names are not as important as the types of work they do and the loads they can bear. The three basic types are (1) flat, (2) spiral, and (3) helical.

Flat Springs

Flat springs include various forms of elliptic or leaf springs (fig. 11-7, A [1] and [2]), made up of flat or

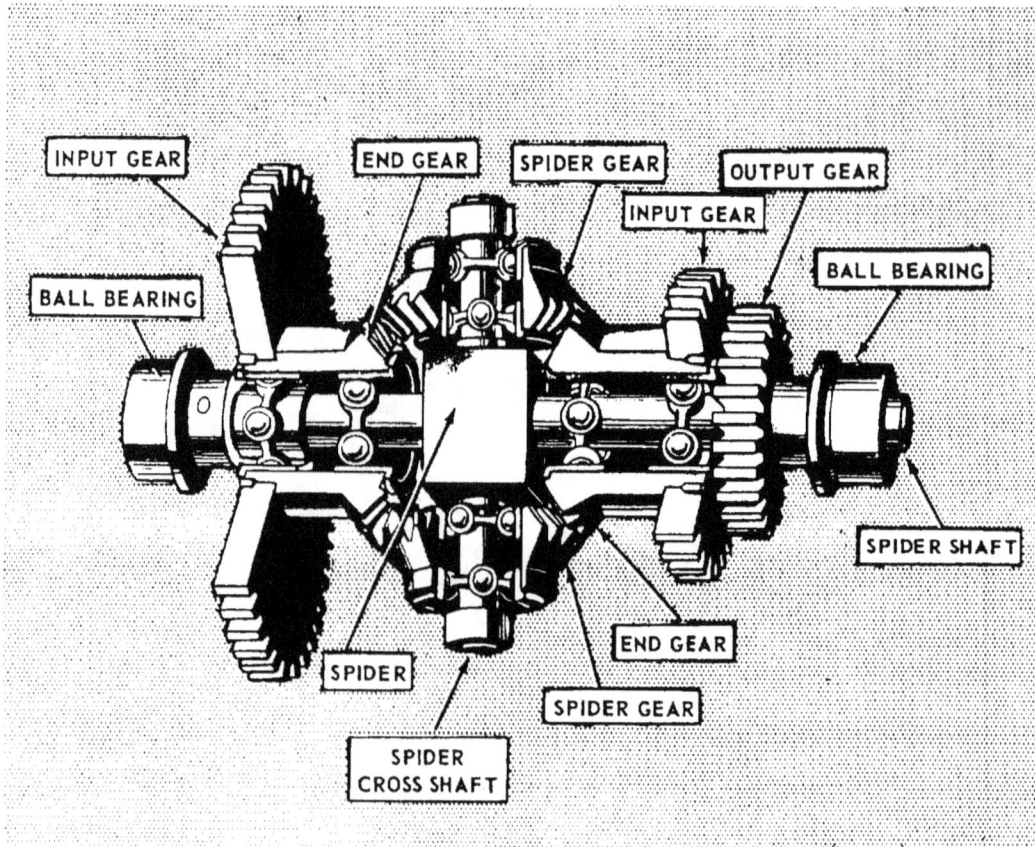

Figure 11-8.-Bevel gear differential.

slightly curved bars, plates, or leaves. They also include special flat springs (fig. 11-7, A [3]), made from a flat strip or bar formed into whatever shape or design best suited for a specific position and purpose.

Spiral Springs

Spiral springs are sometimes called clock, power (1 1-7, B), or coil springs. A well-known example is a watch or clock spring; after you wind (tighten) it, it gradually unwinds and releases power. Although other names for these springs arc based on good authority, we call them "spiral" in this text to avoid confusion.

Helical Springs

Helical springs, also often called spiral (fig. 11-7, D), are probably the most common type of spring. They may be used in compression (fig. 11-7, D [1]), extension or tension (fig. 11-7, D [2]), or torsion (fig. 11-7, D [3]). A spring used in compression tends to shorten in action,

while a tension spring lengthens in action. Torsion springs, which transmit a twist instead of a direct pull, operate by a coiling or an uncoiling action.

In addition to straight helical springs, cone, double-cone, keg, and volute springs are classified as helical. These types of springs are usually used in compression. A cone spring (11-7, D [4]), often called a valve spring because it is frequently used in valves, is formed by wire being wound on a tapered mandrel instead of a straight one. A double cone spring (not illustrated) consists of two cones joined at the small ends, and a keg spring (not illustrated) consists of two cone springs joined at their large ends.

Volute springs (fig. 11-7, D [5]) are conical springs made from a flat bar that is wound so that each coil partially overlaps the adjacent one. The width (and thickness) of the material gives it great strength or resistance.

You can press a conical spring flat so that it requires little space, and it is not likely to buckle sidewise.

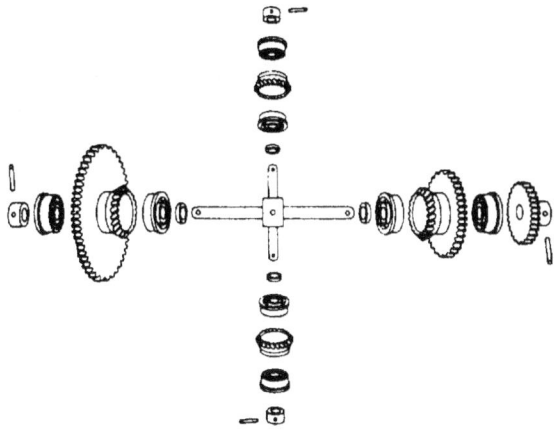

Figure 11-9.-Exploded view of differential gear system.

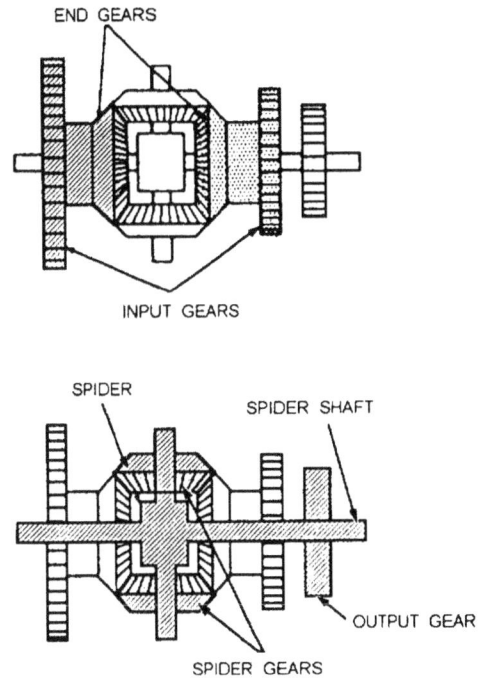

Figure 11-10.-The differential. End gears and spider arrangement.

Other Types of Springs

Torsion bars (fig. 11-7, C) are straight bars that are acted on by torsion (twisting force). The bars may be circular or rectangular in cross section. They also may be tube shaped; other shapes are uncommon.

A special type of spring is a ring spring or disc spring (not illustrated). It is made of several metal rings or discs that overlap each other.

THE GEAR DIFFERENTIAL

A gear differential is a mechanism that is capable of adding and subtracting mechanically. To be more precise, we should say that it adds the total revolutions of two shafts. It also subtracts the total revolutions of one shaft from the total revolutions of another shaft—and delivers the answer by a third shaft. The gear differential will continuously and accurately add or subtract any number of revolutions. It will produce a continuous series of answers as the inputs change.

Figure 11-8 is a cutaway drawing of a bevel gear differential showing all of its parts and how they relate to each other. Grouped around the center of the mechanism are four bevel gears meshed together. The two bevel gears on either side are "end gears." The two bevel gears above and below are "spider gears." The long shaft running through the end gears and the three spur gears is the "spider shaft." The short shaft running through the spider gears together with the spider gears themselves make up the "spider."

Each spider gear and end gear is bearing-mounted on its shaft and is free to rotate. The spider shaft connects with the spider cross shaft at the center block where they intersect. The ends of the spider shaft are secured in flanges or hangers. The spider cross shaft and the spider shaft are also bearing-mounted and are free to rotate on their axis. Therefore, since the two shafts are rigidly connected, the spider (consisting of the spider cross shaft and the spider gears) must tumble, or spin, on the axis of the spider shaft.

The three spur gears, shown in figure 11-8, are used to connect the two end gears and the spider shaft to other mechanisms. They may be of any convenient size. Each of the two input spur gears is attached to an end gear. An input gear and an end gear together are called a "side" of a differential. The third spur gear is the output gear, as designated in figure 11-8. This is the only gear pinned to the spider shaft. All the other differential gears, both bevel and spur, are bearing-mounted.

Figure 11-9 is an exploded view of a gear differential showing each of its individual parts. Figure 11-10 is a schematic sketch showing the relationship of the principle parts. For the present we will assume that the two sides of the gear system are the inputs and the gear on the spider shaft is the output. Later we will show that any of these three gears can be either an input or an output.

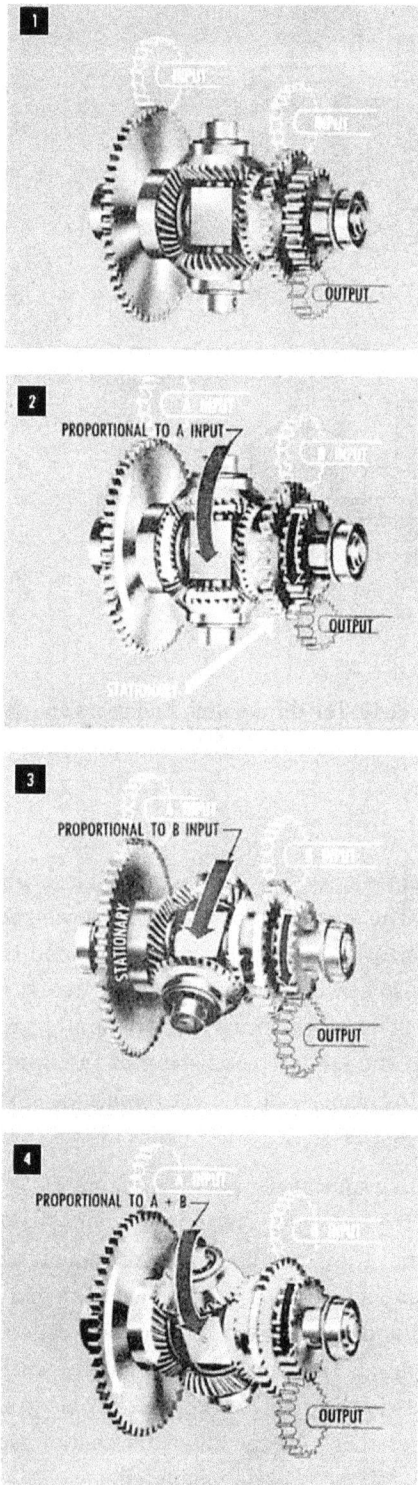

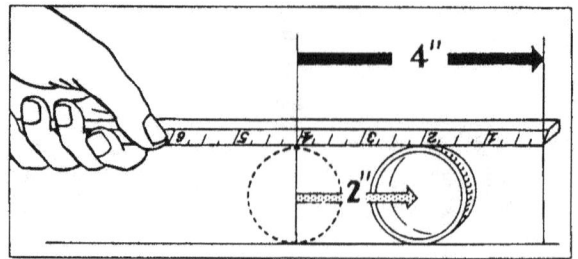

Figure 11-12.—The spider makes only half as many revolutions.

Figure 11-11.—How a differential works.

gears, turning the spider shaft several revolutions proportional to the sum, or difference, of the revolutions of the end gears.

Suppose the left side of the differential rotates while the other remains stationary, as in block 2 of figure 11-11. The moving end gear will drive the spider in the same direction as the input and, through the spider shaft and output gear, the output shaft. The output shaft will turn several revolutions proportional to the input.

If the right side is not rotated and the left side is held stationary, as in block 3 of figure 11-11, the same thing will happen. If both input sides of the differential turn in the same direction at the same time, the spider will be turned by both at once, as in block 4 of figure 11-11. The output will be proportional to the two inputs. Actually, the spider makes only half as many revolutions as the revolutions of the end gears, because the spider gears are free to roll between the end gears. To understand this better, let's look at figure 11-12. Here a ruler is rolled across the upper side of a cylindrical drinking glass, pushing the glass along a table top. The glass will roll only half as far as the ruler travels. The spider gears in the differential roll against the end gears in exactly the same way. Of course, you can correct the way the gears work by using a 2:1 gear ratio between the gear on the spider shaft and the gear for the output shaft. Very often, for design purposes, this gear ratio will be found to be different.

When two sides of the differential move in opposite directions, the output of the spider shaft is proportional to the difference of the revolutions of the two inputs. That is because the spider gears are free to turn and the two inputs drive them in opposite directions. If the two inputs are equal and opposite, the spider gears will turn, but the spider shaft will not move. If the two inputs turn in opposite directions for an unequal number of revolutions, the spider gears roll on the end gear that makes the lesser number of revolutions. That rotates the spider in the direction of the input making the greater number of revolution. The motion of the spider shaft

Now let's look at figure 11-11. In this hookup the two end gears are positioned by the input shafts, which represent the quantities to be added or subtracted. The spider gears do the actual adding and subtracting. They follow the rotation of the two end

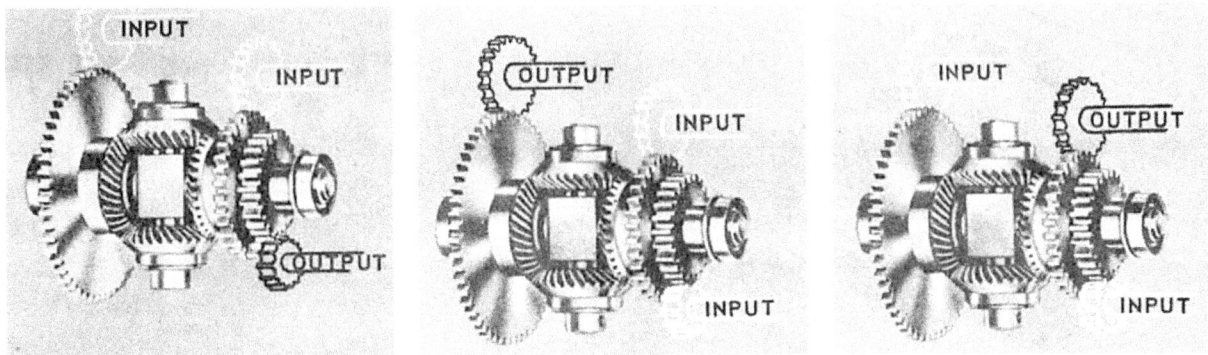

Figure 11-13.—Differential gear hookups.

will be equal to half the difference between the revolutions of the two inputs. A change in the gear ratio to the output shaft can then give us any proportional answer we wish.

We have been describing a hookup wherein the two sides are inputs and the spider shaft is the output. As long as you recognize that the spider follows the end gears for half the sum, or difference, of their revolutions, you don't need to use this type of hookup. You may use the spider shaft as one input and either of the sides as the other. The other side will then become the output. Therefore, you may use three different hookups for any given differential, depending on which is the most convenient mechanically, as shown in figure 11-13.

In chapter 13 of this book, we will describe the use of the differential gear in the automobile. Although this differential is similar in principle, you will see that it is somewhat different in its mechanical makeup.

LINKAGES

A linkage may consist of either one or a combination of the following basic parts:

1. Rod, shaft, or plunger

2. Lever

3. Rocker arm

4. Bell crank

These parts combined will transmit limited rotary or linear motion. To change the direction of a motion, we use cams with the linkage.

Lever-type linkages (fig. 11-14) are used in equipment that you open and close; for instance, valves in electric-hydraulic systems, gates clutches, and clutch-solenoid interlocks. Rocker

arms are merely a variation, or special use, of levers.

Bell cranks primarily transmit motion from a link traveling in one direction to another link moving in a different direction. The bell crank mounts on a fixed

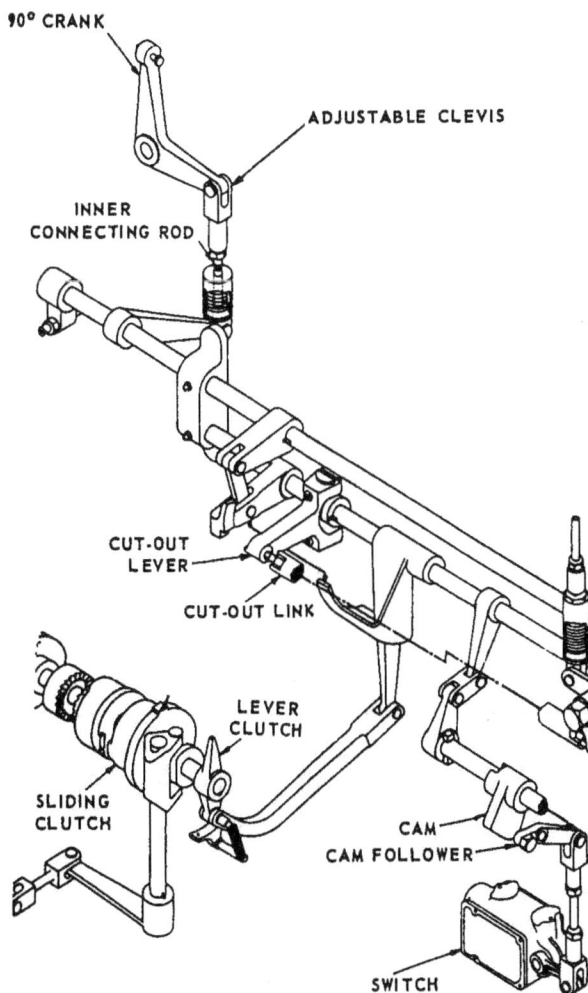

Figure 11-14.—Linkages.

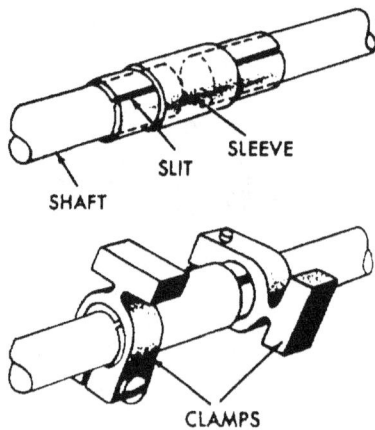

Figure 11-15.-Sleeve coupling.

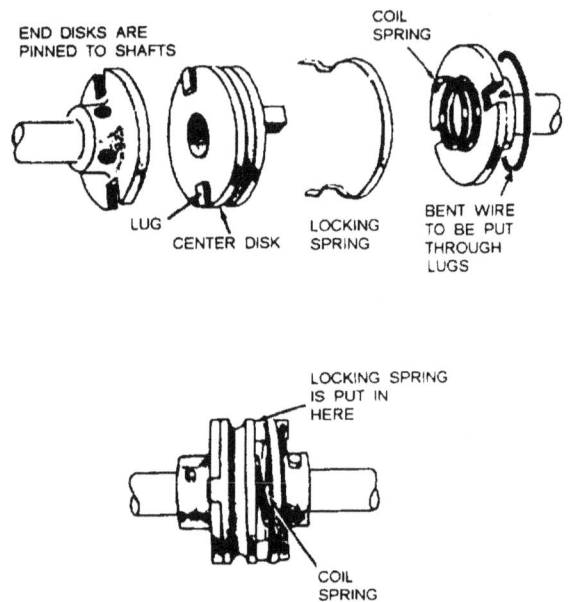

Figure 11-16.-Oldham coupling.

pivot, and the two links connect at two points in different directions from the pivot. By properly locating the connection points, the output links can move in any desired direction.

All linkages require occasional adjustments or repair, particularly when they become worn. To make the proper adjustments, a person must be familiar with the basic parts that constitute a linkage. Adjustments are normally made by lengthening or shortening the rods and shafts by a clevis or turnbuckle.

COUPLINGS

The term *coupling* applies to any device that holds two parts together. Line shafts that make up several shafts of different lengths may be held together by any of several types of shaft couplings.

SLEEVE COUPLING

You may use the sleeve coupling (fig. 11-15) when shafts are closely aligned. It consists of a metal tube slit at each end. The slitted ends enable the clamps to fasten the sleeve securely to the shaft ends. With the clamps tightened, the shafts are held firmly together and turn as one shaft. The sleeve coupling also serves as a convenient device for making adjustments between units. The weight at the opposite end of the clamp from the screw merely offsets the weight of the screw and clamp arms. Distributing the weight evenly reduces the shaft vibration.

OLDHAM COUPLING

The Oldham coupling, named for its inventor, transmits rotary motion between shafts that are parallel but not always in perfect alignment.

An Oldham coupling (fig. 11-16) consists of a pair of disks, one flat and the other hollow. These disks are pinned to the ends of the shafts. A third (center) disk, with a pair of lugs projecting from each face of the disk, fits into the slots between the two end disks and enables one shaft to drive the other shaft. A coil spring, housed within the center of the hollow end disk, forces the center disk against the flat disk. When the coupling is assembled on the shaft ends, a flat lock spring is slipped into the space around the coil spring. The ends of the flat spring are formed so that when they are pushed into the proper place, the ends of the spring push out and lock around the lugs. A lock wire is passed between the holes drilled through the projecting lugs to guard the assembly. The coil spring compensates for any change in shaft length. (Changes in temperature may cause the shaft length to vary.)

The disks, or rings, connecting the shafts allow a small amount of radial play. This play allows a small amount of misalignment of the shafts as they rotate. You can easily connect and disconnect the Oldham type couplings to realign the shafts.

OTHER TYPES OF COUPLINGS

We use four other types of couplings extensively in naval equipment:

1. The fixed (sliding lug) coupling, which is nonadjustable; it does allow for a small amount of misalignment in shafting (fig. 11-17).

2. The flexible coupling (fig. 11-18), which connects two shafts by a metal disk. Two coupling hubs,

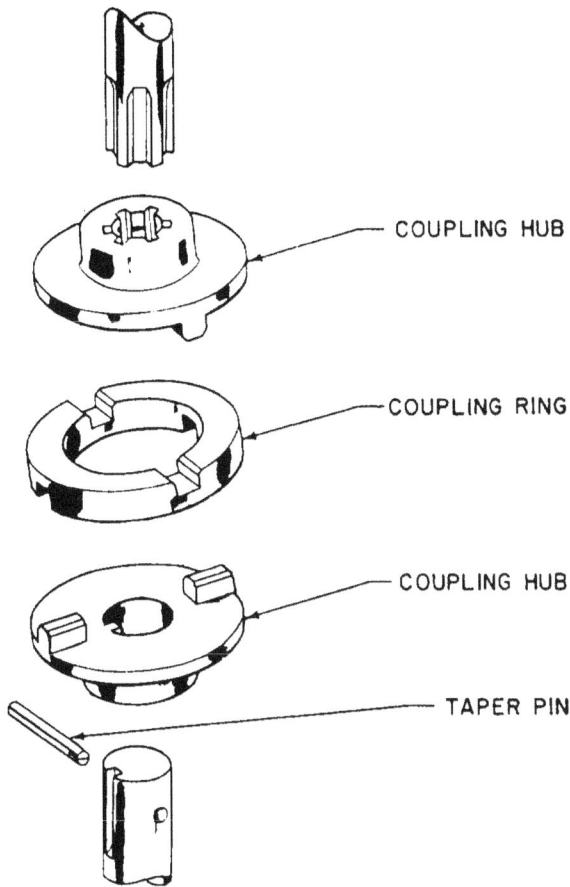

Figure 11-17.-Fixed coupling.

- COUPLING HUB
- COUPLING RING
- COUPLING HUB
- TAPER PIN

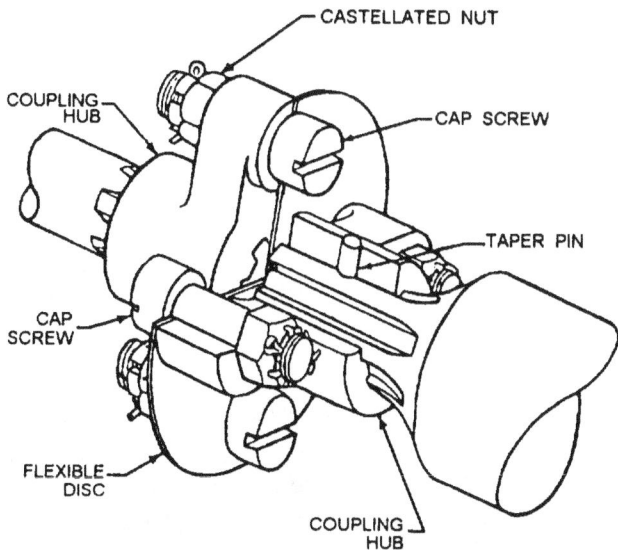

Figure 11-18.-Flexible coupling.

- CASTELLATED NUT
- COUPLING HUB
- CAP SCREW
- TAPER PIN
- CAP SCREW
- FLEXIBLE DISC
- COUPLING HUB

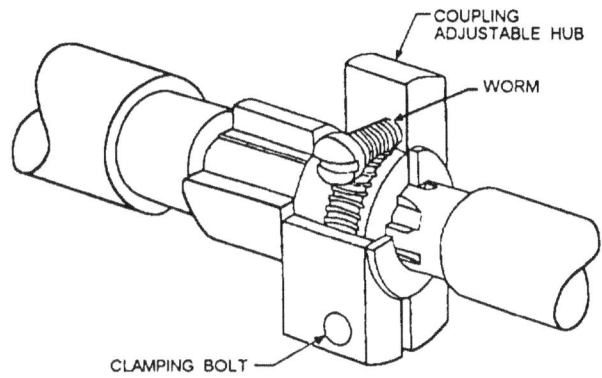

Figure 11-19.-Adjustable (vernier) coupling.

- COUPLING ADJUSTABLE HUB
- WORM
- CLAMPING BOLT

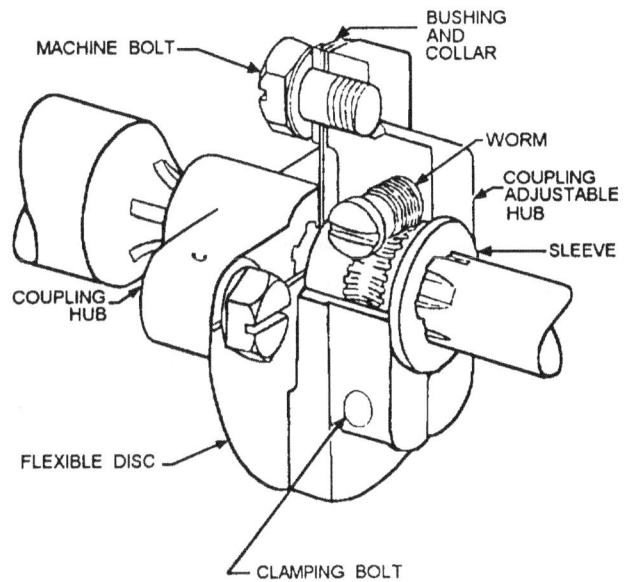

Figure 11-20.-Adjustable flexible (vernier) coupling.

- MACHINE BOLT
- BUSHING AND COLLAR
- WORM
- COUPLING ADJUSTABLE HUB
- SLEEVE
- COUPLING HUB
- FLEXIBLE DISC
- CLAMPING BOLT

3. The adjustable (vernier) coupling, which provides a means of finely adjusting the relationship of two interconnected rotating shafts (fig. 11-19). Loosening a clamping bolt and turning an adjusting worm allows one shaft to rotate while the other remains stationary. After attaining the proper relationship, you retighten the clamping bolt to lock the shafts together again.

4. The adjustable flexible (vernier) coupling (fig. 11-20), which is a combination of the flexible disk coupling and the adjustable (vernier) coupling.

UNIVERSAL JOINT

To couple two shafts in different planes, you need to use a universal joint. Universal joints have various

each splined to its respective shaft, are bolted to the metal disk. The flexible coupling provides a small amount of flexibility to allow for a slight axial misalignment of the shafts.

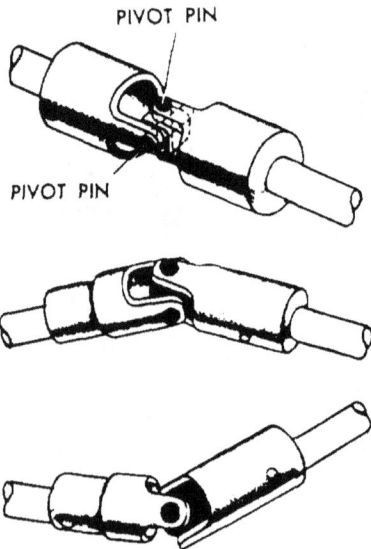

PIVOT PIN

PIVOT PIN

Figure 11-21.-Universal joint (Hooke type).

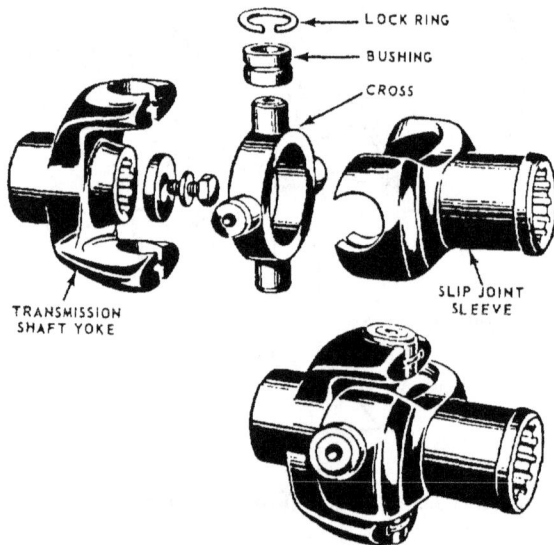

LOCK RING

BUSHING

CROSS

TRANSMISSION SHAFT YOKE

SLIP JOINT SLEEVE

Figure 11-22.-Ring-and-trunnion universal joint.

forms. They are used in nearly all types and classes of machinery. An elementary universal joint, sometimes called a Hooke joint (fig. 11-21), consists of two U-shaped yokes fastened to the ends of the shafts to be connected. Within these yokes is a cross-shaped part that holds the yokes together and allows each yoke to bend, or pivot, in relation to the other. With this arrangement, one shaft can drive the other even though the angle between the two is as great as 25° from alignment.

Figure 11-22 shows a ring-and-trunnion universal joint. It is merely a slight modification of the old Hooke joint. Automobile drive shaft systems use two, and

sometimes three, of these joints. You will read more about these in chapter 13 of this book.

The Bendix-Weiss universal joint (fig. 11-23) provides smoother torque transmission but less structural strength. In this type of joint, four large balls transmit the rotary force, with a smaller ball as a spacer. With the Hooke type universal joint, a whipping motion occurs as the shafts rotate. The amount of whip depends on the degree of shaft misalignment. The Bendix-Weiss joint does not have this disadvantage; it transmits rotary motion with a constant angular velocity. However, this type of joint is both more expensive to manufacture and of less strength than the Hooke type.

CAMS

A cam is a rotating or sliding piece of machinery (as a wheel or a projection on a wheel). A cam transfers motion to a roller moving against its edge or to a pin free to move in a groove on its face. A cam may also receive motion from such a roller or pin. Some cams do not move at all, but cause a change of motion in the contacting part. Cams are not ordinarily used to transmit power in the sense that gear trains are used. They are used to modify mechanical movement, the power for which is furnished through other means. They may control other mechanical units, or they may synchronize or lock together two or more engaging units.

Cams are of many shapes and sizes and are widely used in machines and machine tools (fig. 11-24). We classify cams as

1. radial or plate cams,

2. cylindrical or barrel cams, and

3. pivoted beams.

A similar type of cam includes drum or barrel cams, edge cams, and face cams.

The drum or barrel cam has a path cut around its outside edge in which the roller or follower fits. It imparts a to-and-from motion to a slide or lever in a plane parallel to the axis of the cam. Sometimes we build these cams upon a plain drum with cam plates attached.

Plate cams are used in 5"/38 and 3"/50 guns to open the breechblock during counter-recoil.

Edge or peripheral cams, also called disc cams, operate a mechanism in one direction only. They rely on gravity or a spring to hold the roller in contact with the edge of the cam. The shape of the cam suits the action required.

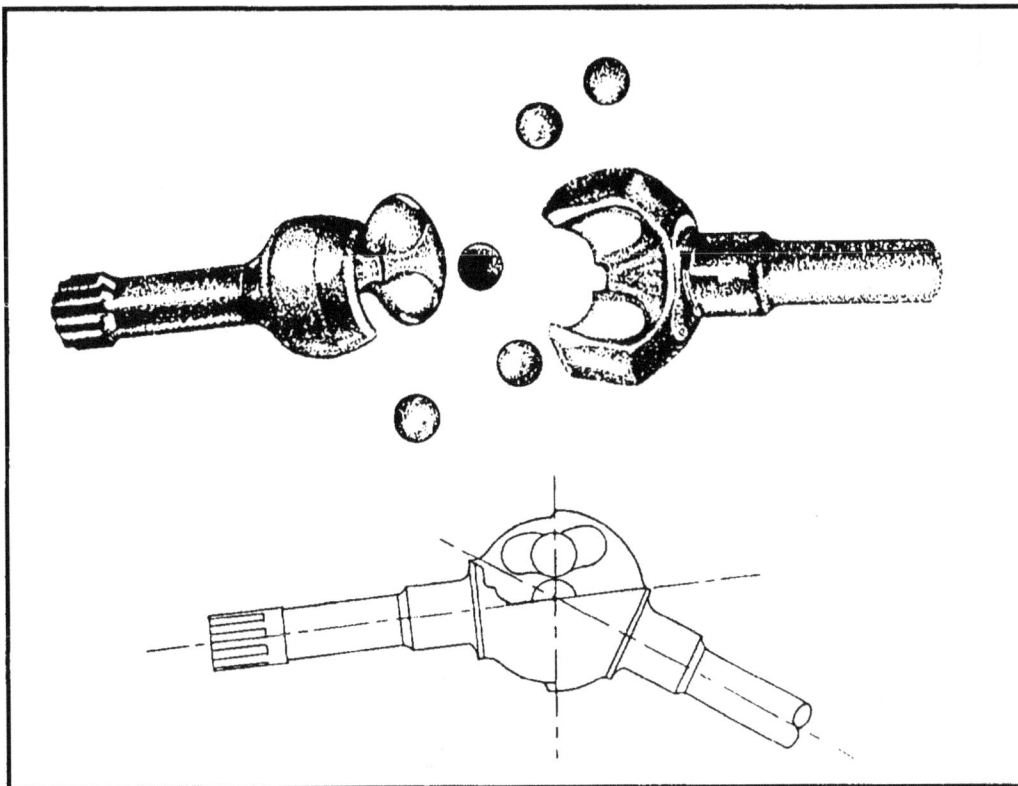

Figure 11-23.-Bendix-Weiss universal joint.

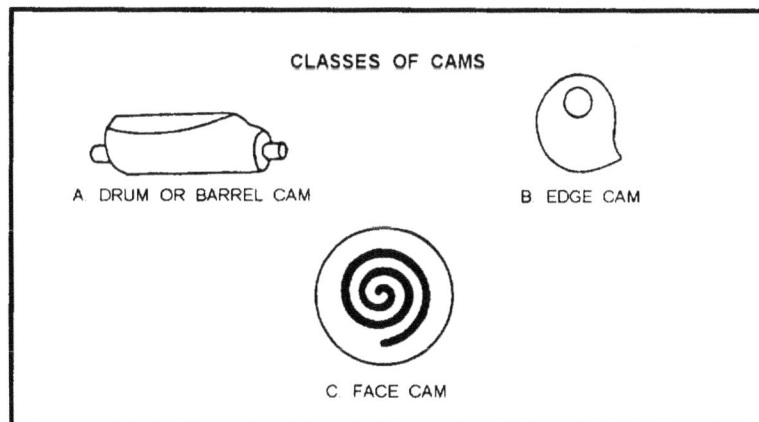

Figure 11-24.-Classes of cams.

Face cams have a groove or slot cut in the face to provide a path for the roller. They operate a lever or other mechanism positively in both directions. The roller is guided by the sides of the slot. Such a groove can be seen on top of the bolt of the Browning .30-caliber machine gun or in fire control cams. The shape of the groove determines the name of the cam, for example, the square cam.

CLUTCHES

A clutch is a form of a coupling. It is designed to connect or disconnect a driving and a driven part as a

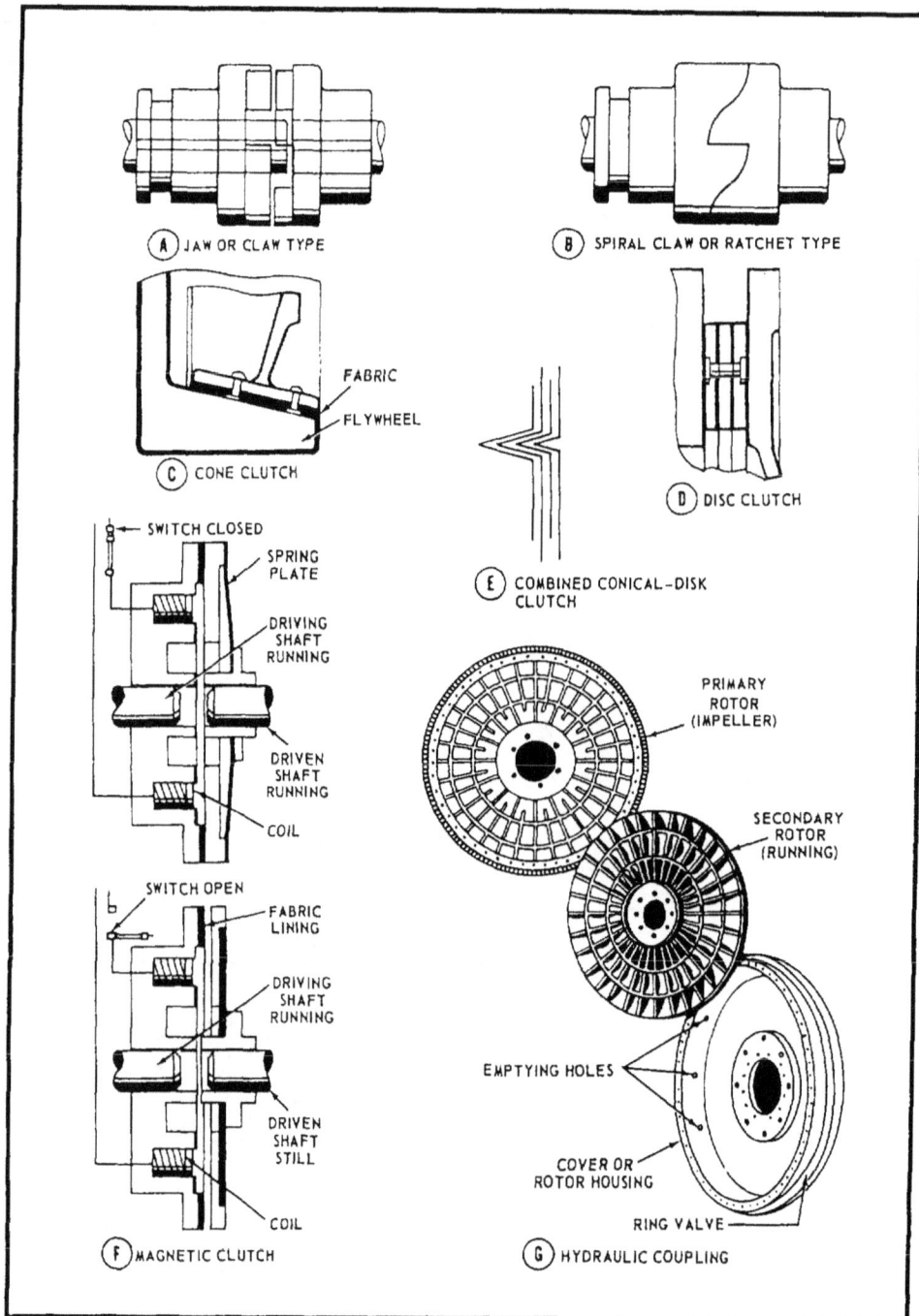

Figure 11-25.-Types of clutches.

means of stopping or starting the driven part. There are two general classes of clutches: positive clutches and friction clutches.

Positive clutches have teeth that interlock. The simplest is the jaw or claw type (fig. 11-25, A), usable only at low speeds. The teeth of the spiral claw or ratchet type (fig. 11-25, B) interlock only one way—they cannot be reversed. An example of this type of clutch is

that seen in bicycles. It engages the rear sprocket with the rear wheel when the pedals are pushed forward and lets the rear wheel revolve freely when the pedals are stopped.

The object of a friction clutch is to connect a rotating member to one that is stationary, to bring it up to speed, and to transmit power with a minimum of slippage. Figure 11-25, C, shows a cone clutch commonly used

11-14

in motor trucks. Friction clutches may be single-cone or double-cone. Figure 11-25, D, shows a disc clutch, also used in autos. A disc clutch also may have several plates (multiple-disc clutch). In a series of discs, each driven disc is located between two driving discs. You may have had experience with a multiple-disc clutch on your car.

The Hele-Shaw clutch is a combined conical-disc clutch (fig. 11-25, E). Its groove permits cooling and circulation of oil. Single-disc clutches are frequently dry clutches (no lubrication); multiple-disc clutches may be dry or wet (either lubricated or operated with oil).

Magnetic clutches are a recent development in which the friction surfaces are brought together by magnetic force when the electricity is turned on (fig. 11-25, F). The induction clutch transmits power without contact between the driving and driven parts.

The way pressure is applied to the rim block, split ring, band, or roller determines the names of expanding clutches or rim clutches. In one type of expanding clutch, right- and left-hand screws expand as a sliding sleeve moves along a shaft and expands the band against the rim. The centrifugal clutch is a special application of a block clutch.

Machines containing heavy parts to be moved, such as a rolling mill, use oil clutches. The grip of the coil causes great friction when it is thrust onto a cone on the driving shaft. Yet the clutch is very sensitive to control.

Diesel engines and transportation equipment use pneumatic and hydraulic clutches. Hydraulic couplings (fig, 11-25, G), which also serve as clutches, are used in the hydraulic A-end of electric-hydraulic gun drives.

SUMMARY

In this chapter we discussed the following elements and mechanisms used in naval machinery:

Two types of bearings are used in naval machinery: sliding and antifrictional.

Springs are another element used in machinery. Springs can be twisted, pulled, or stretched by force and can return to their original shape when the force is released.

One basic mechanism of machines is the gear differential. A gear differential is a mechanism that is capable of adding and subtracting mechanically. Other basic mechanisms include linkages, couplings, cams and cam followers, and clutches.

CHAPTER 12

INTERNAL COMBUSTION ENGINE

CHAPTER LEARNING OBJECTIVES

Upon completion of this chapter, you should be able to do the following:

- *Explain the principles of a combustion engine.*

- *Explain the process of an engine cycle.*

- *State the classifications of engines.*

- *Discuss the construction of an engine.*

- *List the auxiliary assemblies of an engine.*

The automobile is a familiar object to all of us. The engine that moves it is one of the most fascinating and talked about of all the complex machines we use today. In this chapter we will explain briefly some of the operational principles and basic mechanisms of this machine. As you study its operation and construction, notice that it consists of many of the devices and basic mechanisms covered earlier in this book.

COMBUSTION ENGINE

We define an engine simply as a machine that converts heat energy to mechanical energy. The engine does this through either internal or external combustion.

Combustion is the act of burning. Internal means inside or enclosed. Thus, in internal combustion engines, the burning of fuel takes place inside the engine; that is, burning takes place within the same cylinder that produces energy to turn the crankshaft. In external combustion engines, such as steam engines, the burning of fuel takes place outside the engine. Figure 12-1 shows, in the simplified form, an external and an internal combustion engine.

The external combustion engine contains a boiler that holds water. Heat applied to the boiler causes the water to boil, which, in turn, produces steam. The steam passes into the engine cylinder under pressure and forces the piston to move downward. With the internal

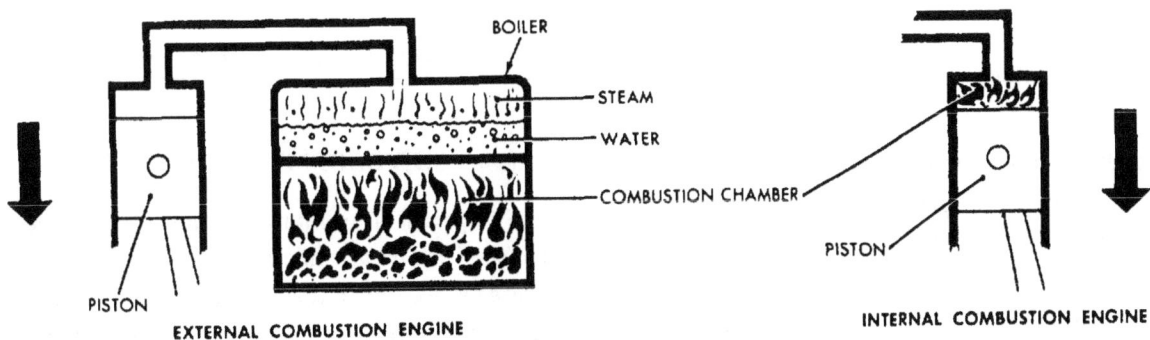

Figure 12-1.-Simple external and internal combustion engine.

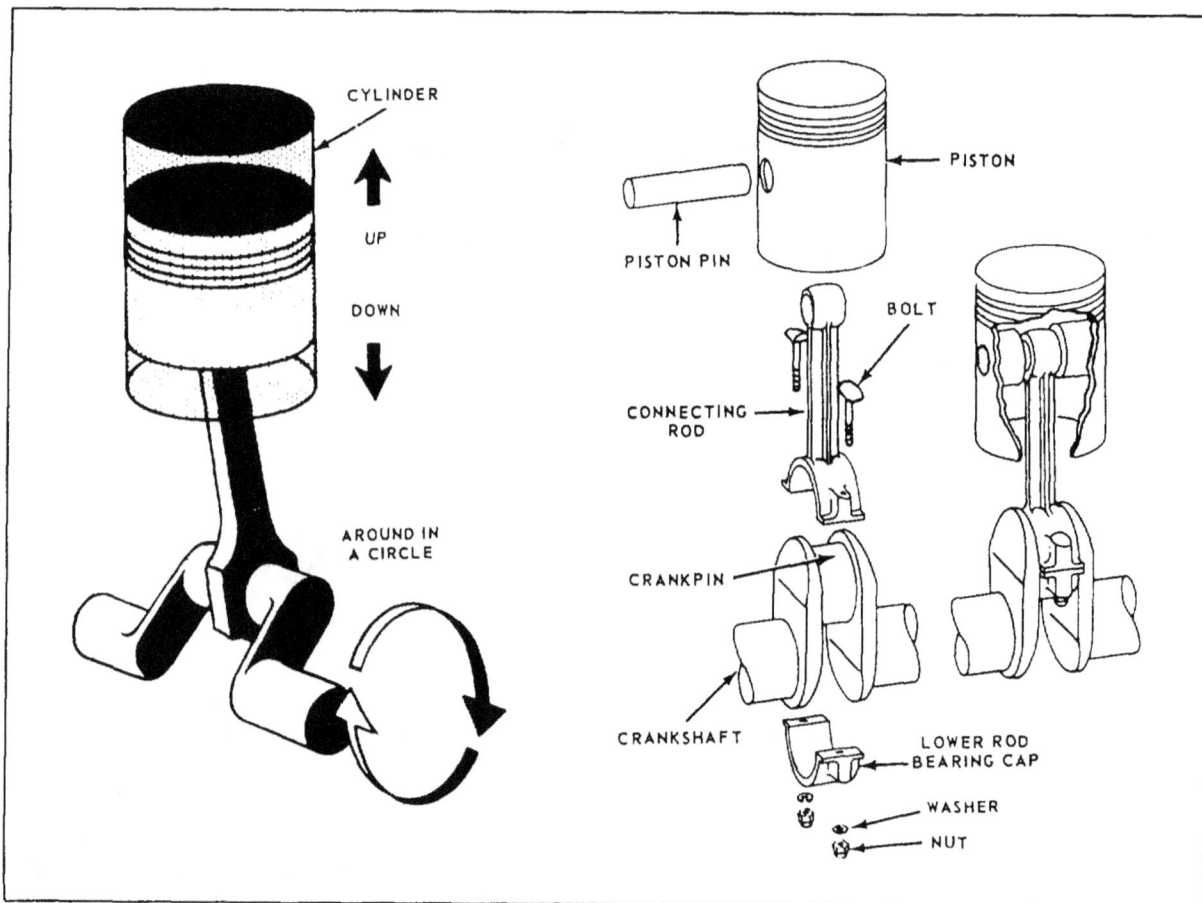

Figure 12-2.-Cylinder, piston, connecting rod, and crankshaft for a one-cylinder engine.

combustion engine, the combustion takes place inside the cylinder and is directly responsible for forcing the piston to move downward.

The change of heat energy to mechanical energy by the engine is based on a fundamental law of physics. It states that gas will expand upon the application of heat. The law also states that the compression of gas will increase its temperature. If the gas is confined with no outlet for expansion, the application of heat will increase the pressure of the gas (as it does in an automotive cylinder). In an engine, this pressure acts against the head of a piston, causing it to move downward.

As you know, the piston moves up and down in the cylinder. The up-and-down motion is known as reciprocating motion. This reciprocating motion (straight line motion) must change to rotary motion (turning motion) to turn the wheels of a vehicle. A crank and a connecting rod change this reciprocating motion to rotary motion.

All internal combustion engines, whether gasoline or diesel, are basically the same. They all rely on three elements: air, fuel, and ignition.

Fuel contains potential energy for operating the engine; air contains the oxygen necessary for combustion; and ignition starts combustion. All are fundamental, and the engine will not operate without any one of them. Any discussion of engines must be based on these three elements and the steps and mechanisms involved in delivering them to the combustion chamber at the proper time.

DEVELOPMENT OF POWER

The power of an internal combustion engine comes from the burning of a mixture of fuel and air in a small, enclosed space. When this mixture burns, it expands; the push or pressure created then moves the piston, thereby cranking the engine. This movement is sent back to the wheels to drive the vehicle.

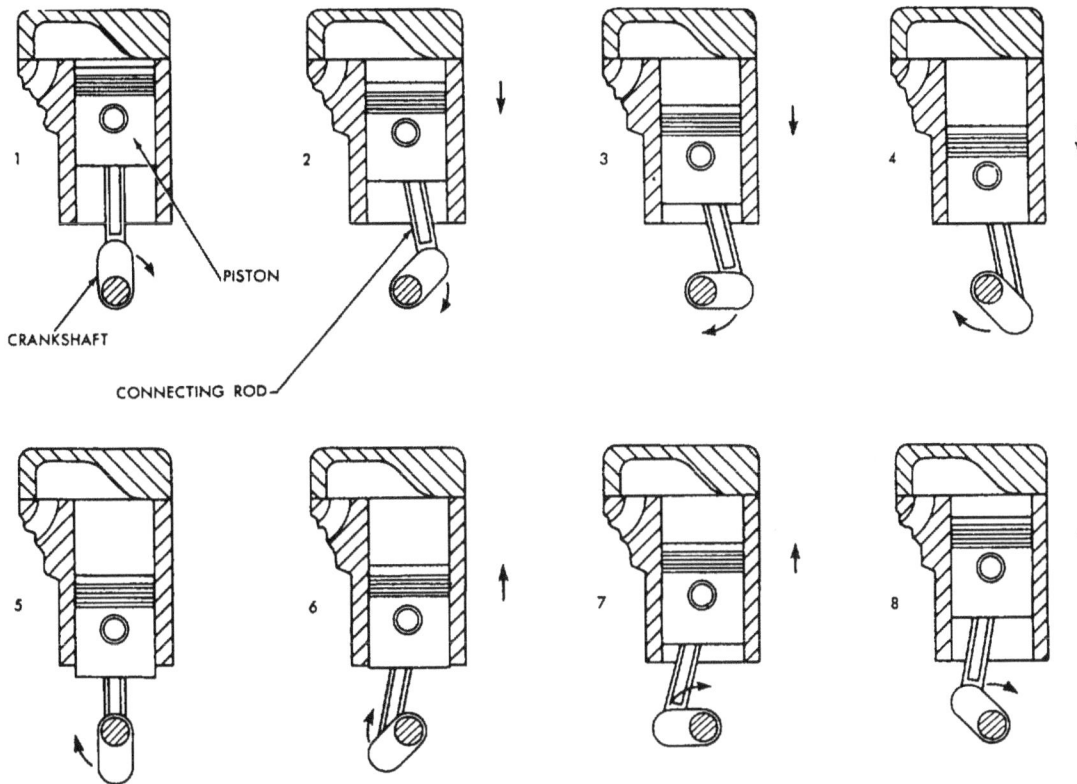

Figure 12-3.-Relationship of piston, connecting rod, and crank on crankshaft as crankshaft turns one revolution.

Since similar action occurs in all cylinders of an engine, we will describe the use one cylinder in the development of power. The one-cylinder engine consists of four basic parts: cylinder, piston, connecting rod, and crankshaft (shown in fig. 12-2).

The cylinder, which is similar to a tall metal can, is closed at one end. Inside the cylinder is the piston, a movable metal plug that fits snugly into the cylinder, but can still slide up and down easily. This up-and-down movement, produced by the burning of fuel in the cylinder, results in the production of power from the engine.

You have already learned that the up-and-down movement is called reciprocating motion. This motion must be changed to rotary motion to rotate the wheels or tracks of vehicles. This change is accomplished by a crank on the crankshaft and a connecting rod between the piston and the crank.

The crankshaft is a shaft with an offset portion-the crank— that describes a circle as the shaft rotates. The top end of the connecting rod connects to the piston and must therefore go up and down. Since the lower end of the connecting rod attaches to the crankshaft, it moves in a circle; however it also moves up and down.

When the piston of the engine slides downward because of the pressure of the expanding gases in the cylinder, the upper end of the connecting rod moves downward with the piston in a straight line. The lower end of the connecting rod moves down and in a circular motion at the same time. This moves the crank; in turn, the crank rotates the shaft. This rotation is the desired result. So remember, the crankshaft and connecting rod combination is a mechanism for changing straight-line, up-and-down motion to circular, or rotary, motion.

BASIC ENGINE STROKES

Each movement of the piston from top to bottom or from bottom to top is called a stroke. The piston takes two strokes (an upstroke and a downstroke) as the crankshaft makes one complete revolution. When the piston is at the top of a stroke, it is said to be at top dead center. When the piston is at the bottom of a stroke, it is said to be at bottom dead center. These positions are rock positions, which we will discuss further in this chapter under "Timing." See figure 12-3 and figure 12-7.

The basic engine you have studied so far has had no provisions for getting the fuel-air mixture into the cylinder or burned gases out of the cylinder. The

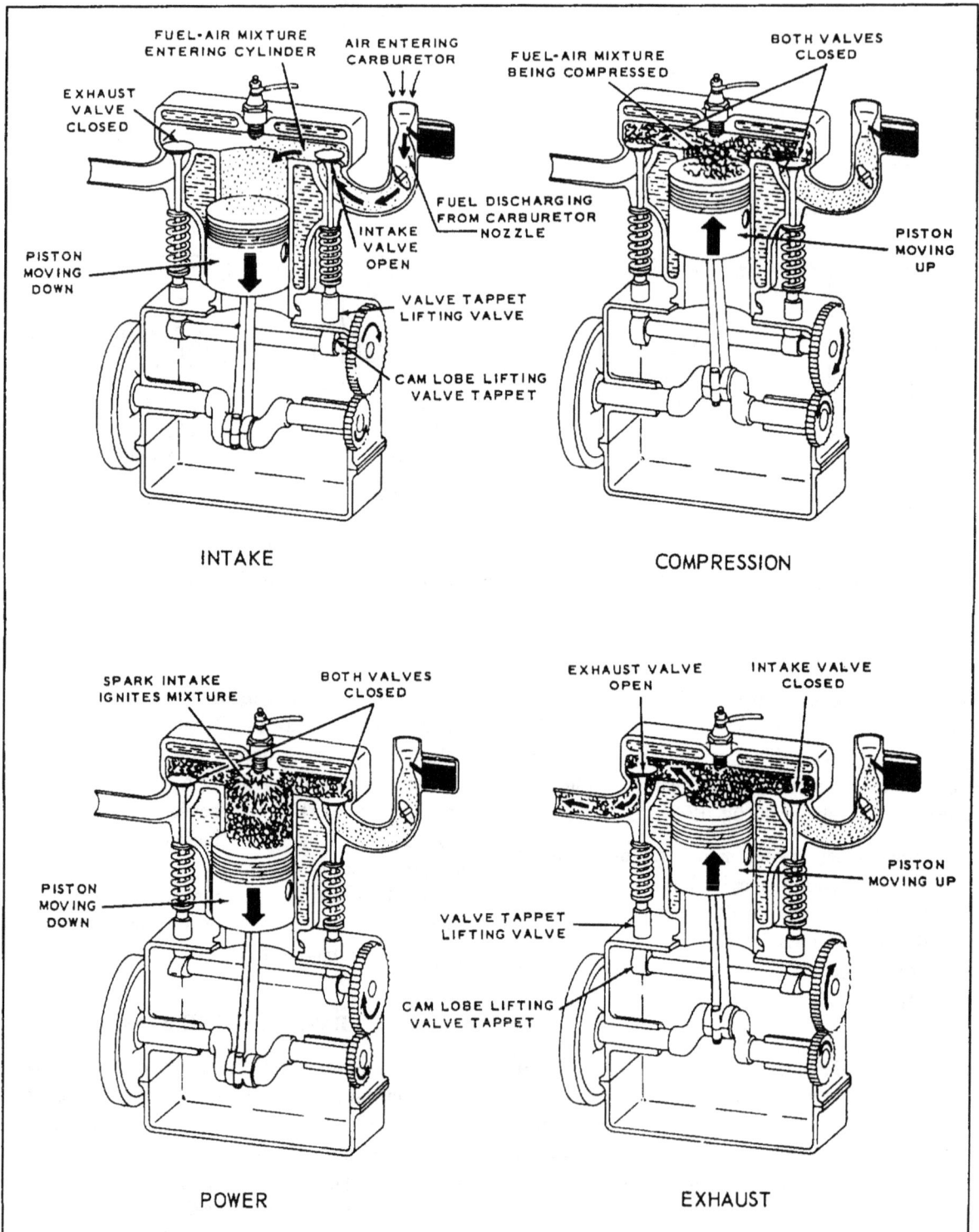

Figure 12-4.-Four-stroke cycle in a gasoline engine.

enclosed end of a cylinder has two openings. One of the openings, or ports, permits the mixture of air and fuel to enter, and the other port permits the burned gases to escape from the cylinder. The two ports have valves assembled in them. These valves, actuated by the camshaft, close off either one or the other of the ports, or both of them, during various stages of engine operation. One of the valves, called the intake valve, opens to admit a mixture of fuel and air into the cylinder. The other valve, called the exhaust valve, opens to allow the escape of burned gases after the fuel-and-air mixture has burned. Later you will learn more about how these valves and their mechanisms operate.

The following paragraphs explain the sequence of actions that takes place within the engine cylinder: the intake stroke, the compression stroke, the power stroke, and the exhaust stroke. Since these strokes are easy to identify in the operation of a four-cycle engine, that engine is used in the description. This type of engine is called a four-stroke-Otto-cycle engine, named after Dr. N. A. Otto who, in 1876, first applied the principle of this engine.

INTAKE STROKE

The first stroke in the sequence is the intake stroke (fig. 12-4). During this stroke, the piston is moving downward and the intake valve is open. This downward movement of the piston produces a partial vacuum in the cylinder, and air and fuel rush into the cylinder past the open intake valve. This action produces a result similar to that which occurs when you drink through a straw. You produce a partial vacuum in your mouth, and the liquid moves up through the straw to fill the vacuum.

COMPRESSION STROKE

When the piston reaches bottom dead center at the end of the intake stroke (and is therefore at the bottom of the cylinder) the intake valve closes and seals the upper end of the cylinder. As the crankshaft continues to rotate, it pushes the connecting rod up against the piston. The piston then moves upward and compresses the combustible mixture in the cylinder. This action is known as the compression stroke (fig. 12-4). In gasoline engines, the mixture is compressed to about one-eighth of its original volume. (In a diesel engine the mixture may be compressed to as little as one-sixteenth of its original volume.) This compression of the air-fuel mixture increases the pressure within the cylinder. Compressing the mixture in this way makes it more

combustible; not only does the pressure in the cylinder go up, but the temperature of the mixture also increases.

POWER STROKE

As the piston reaches top dead center at the end of the compression stroke (and is therefore at the top of the cylinder), the ignition system produces an electric spark. The spark sets fire to the fuel-air mixture. In burning, the mixture gets very hot and expands in all directions. The pressure rises to about 600 to 700 pounds per square inch. Since the piston is the only part that can move, the force produced by the expanding gases forces the piston down. This force, or thrust, is carried through the connecting rod to the crankpin on the crankshaft. The crankshaft is given a powerful twist. This is known as the power stroke (fig. 12-4). This turning effort, rapidly repeated in the engine and carried through gears and shafts, will turn the wheels of a vehicle and cause it to move along the highway.

EXHAUST STROKE

After the fuel-air mixture has burned, it must be cleared from the cylinder. Therefore, the exhaust valve opens as the power stroke is finished and the piston starts back up on the exhaust stroke (fig. 12-4). The piston forces the burned gases of the cylinder past the open exhaust valve. The four strokes (intake, compression, power, and exhaust) are continuously repeated as the engine runs.

ENGINE CYCLES

Now, with the basic knowledge you have of the parts and the four strokes of the engine, let us see what happens during the actual running of the engine. To produce sustained power, an engine must repeatedly complete one series of the four strokes: intake, compression, power, and exhaust. One completion of this series of strokes is known as a cycle.

Most engines of today operate on four-stroke cycles, although we use the term *four-cycle engines* to refer to them. The term actually refers to the four strokes of the piston, two up and two down, not the number of cycles completed. For the engine to operate, the piston continually repeats the four-stroke cycle.

TWO-CYCLE ENGINE

In the two-cycle engine, the entire series of strokes (intake, compression, power, and exhaust) takes place in two piston strokes.

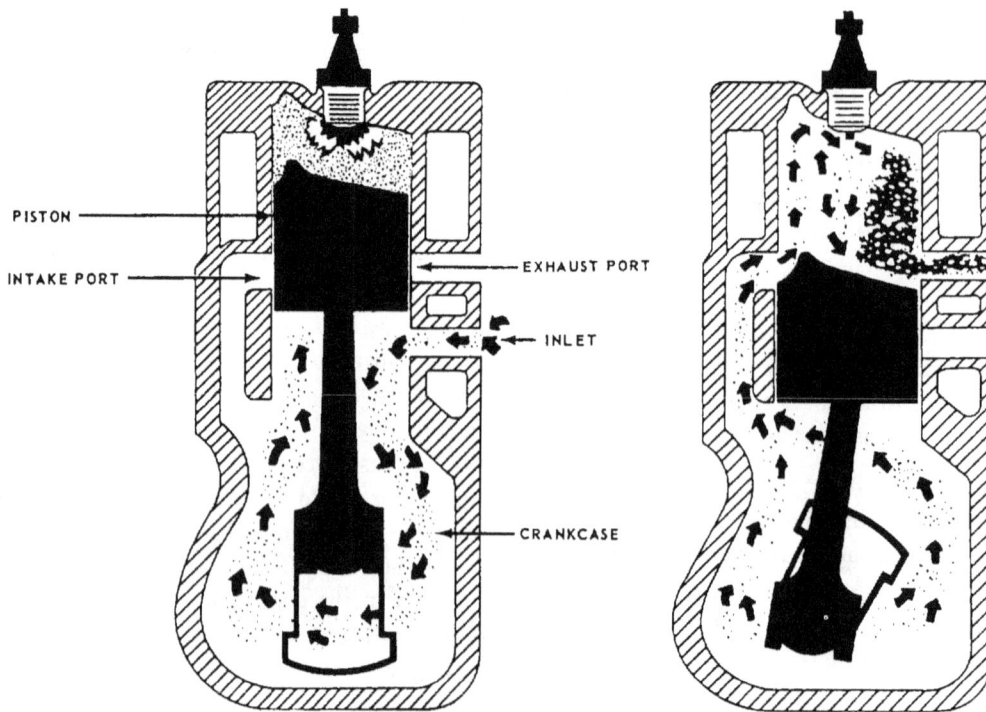

Figure 12-5.-Events in a two-cycle, internal combustion engine.

A two-cycle engine is shown in figure 12-5. Every other stroke in this engine is a power stroke. Each time the piston moves down, it is on the power stroke. Intake, compression, power, and exhaust still take place; but they are completed in just two strokes. Figure 12-5 shows that the intake and exhaust ports are cut into the cylinder wall instead of at the top of the combustion chamber as in the four-cycle engine. As the piston moves down on its power stroke, it first uncovers the exhaust port to let burned gases escape and then uncovers the intake port to allow a new fuel-air mixture to enter the combustion chamber. Then on the upward stroke, the piston covers both ports and, at the same time, compresses the new mixture in preparation for ignition and another power stroke.

In the engine shown in figure 12-5, the piston is shaped so that the incoming fuel-air mixture is directed upward, thereby sweeping out ahead of it the burned exhaust gases. Also, there is an inlet into the crankcase through which the fuel-air mixture passes before it enters the cylinder. This inlet is opened as the piston moves upward, but it is sealed as the piston moves downward on the power stroke. The downward moving piston slightly compresses the mixture in the crankcase. That gives the mixture enough pressure to pass rapidly through the intake port as the piston clears this port. This action improves the sweeping-out, or scavenging, effect of the mixture as it enters and clears the burned gases from the cylinder through the exhaust port.

FOUR-CYCLE VERSUS TWO-CYCLE ENGINES

You have probably noted that the two-cycle engine produces a power stroke every crankshaft revolution; the four-cycle engine requires two crankshaft revolutions for each power stroke. It might appear that the two-cycle engine could produce twice as much power as the four-cycle engine of the same size, operating at the same speed. However, that is not true. With the two-cycle engine, some of the power is used to drive the blower that forces the air-fuel charge into the cylinder under pressure. Also, the burned gases are not cleared from the cylinder. Additionally, because of the much shorter period the intake port is open (compared to the period the intake valve in a four-stroke-cycle is open), a smaller amount of fuel-air mixture is admitted. Hence, with less fuel-air mixture, less power per power stroke is produced compared to the power produced in a four-stroke cycle engine of like size operating at the same speed and under the same conditions. To increase the amount of fuel-air mixture, we use auxiliary devices with the two-stroke engine to ensure delivery of greater amounts of fuel-air mixture into the cylinder.

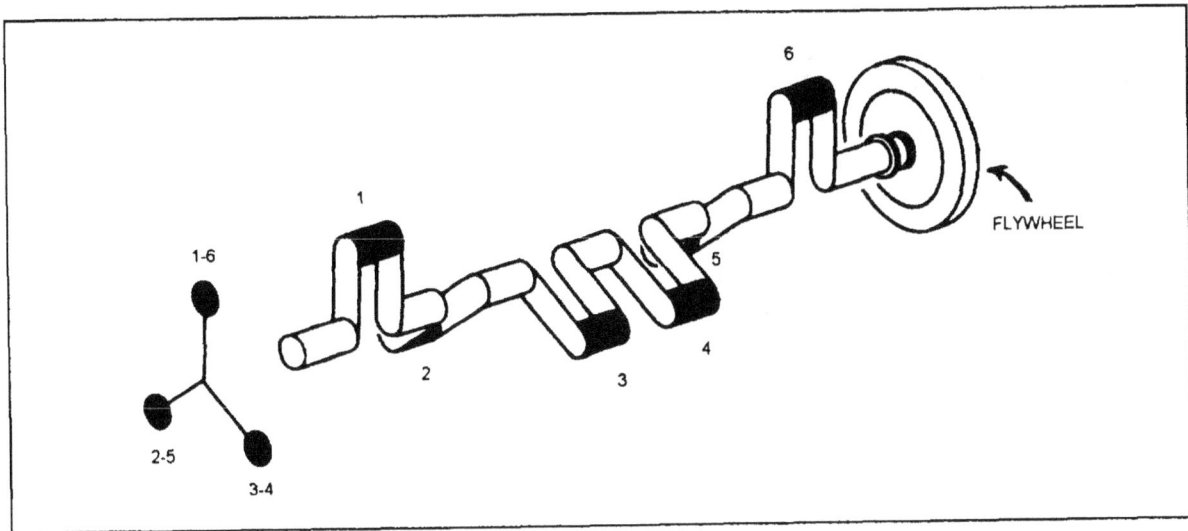

Figure 12-6.-Crankshaft for a six-cylinder engine.

MULTIPLE-CYLINDER ENGINES

The discussion so far in this chapter has concerned a single-cylinder engine. A single cylinder provides only one power impulse every two crankshaft revolutions in a four-cycle engine. It delivers power only one-fourth of the time. To provide for a more continuous flow of power, modem engines use four, six, eight, or more cylinders. The same series of cycles take place in each cylinder.

In a four-stroke cycle, six-cylinder engine, for example, the cranks on the crankshaft are set 120 degrees apart. The cranks for cylinders 1 and 6, 2 and 5, and 3 and 4 are in line with each other (fig. 12-6). The cylinders fire or deliver the power strokes in the following order: 1-5-3-6-2-4. Thus, the power strokes follow each other so closely that a continuous and even delivery of power goes to the crankshaft.

TIMING

In a gasoline engine, the valves must open and close at the proper times with regard to piston position and stroke. In addition, the ignition system must produce the sparks at the proper time so that the power strokes can start. Both valve and ignition system action must be properly timed if good engine performance is to be obtained.

Valve timing refers to the exact times in the engine cycle that the valves trap the mixture and then allow the burned gases to escape. The valves must open and close so that they are constantly in step with the piston movement of the cylinder they control. The position of

the valves is determined by the camshaft; the position of the piston is determined by the crankshaft. Correct valve timing is obtained by providing the proper relationship between the camshaft and the crankshaft.

When the piston is at top dead center, the crankshaft can move 15° to 20° without causing the piston to move up and down any noticeable distance. This is one of the two rock positions (fig. 12-7) of the piston. When the piston moves up on the exhaust stroke, considerable momentum is given to the exhaust gases as they pass out through the exhaust valve port. If the exhaust valve closes at top dead center, a small amount of the gases

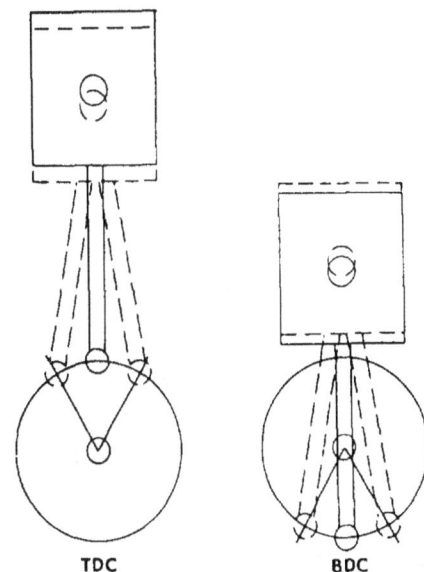

Figure 12-7.-Rock position.

will be trapped and will dilute the incoming fuel-air mixture when the intake valves open. Since the piston has little downward movement while in the rock position, the exhaust valve can remain open during this period and thereby permit a more complete scavenging of the exhaust gases.

Ignition timing refers to the timing of the sparks at the spark plug gap with relation to the piston position during the compression and power strokes. The ignition system is timed so that the sparks occurs before the piston reaches top dead center on the compression stroke. That gives the mixture enough time to ignite and start burning. If this time were not provided, that is, if the spark occurred at or after the piston reached top dead center, then the pressure increase would not keep pace with the piston movement.

At higher speeds, there is still less time for the fuel-air mixture to ignite and bum. To make up for this lack of time and thereby avoid power loss, the ignition system includes an advance mechanism that functions on speed.

CLASSIFICATION OF ENGINES

Engines for automotive and construction equipment may be classified in several ways: type of fuel used, type of cooling employed, or valve and cylinder arrangement. They all operate on the internal combustion principle. The application of basic principles of construction to particular needs or systems of manufacture has caused certain designs to be recognized as conventional.

The most common method of classification is based on the type of fuel used; that is, whether the engine burns gasoline or diesel fuel.

GASOLINE ENGINES VERSUS DIESEL ENGINES

Mechanically and in overall appearance, gasoline and diesel engines resemble one another. However, many parts of the diesel engine are designed to be somewhat heavier and stronger to withstand the higher temperatures and pressures the engine generates. The engines differ also in the fuel used, in the method of introducing it into the cylinders, and in how the air-fuel mixture is ignited. In the gasoline engine, we first mix air and fuel in the carburetor. After this mixture is compressed in the cylinders, it is ignited by an electrical spark from the spark plugs. The source of the energy producing the electrical spark may be a storage battery or a high-tension magneto.

The diesel engine has no carburetor. Air alone enters its cylinders, where it is compressed and reaches a high temperature because of compression. The heat of compression ignites the fuel injected into the cylinder and causes the fuel-air mixture to burn. The diesel engine needs no spark plugs; the very contact of the diesel fuel with the hot air in the cylinder causes ignition. In the gasoline engine the heat compression is not enough to ignite the air-fuel mixture; therefore, spark plugs are necessary.

ARRANGEMENT OF CYLINDERS

Engines are also classified according to the arrangement of the cylinders. One classification is the in-line, in which all cylinders are cast in a straight line above the crankshaft, as in most trucks. Another is the V-type, in which two banks of cylinders are mounted in a "V" shape above the crankshaft, as in many passenger vehicles. Another not-so-common arrangement is the horizontally opposed engine whose cylinders mount in two side rows, each opposite a central crankshaft. Buses often have this type of engine.

The cylinders are numbered. The cylinder nearest the front of an in-line engine is numbered 1. The others are numbered 2, 3,4, and so forth, from the front to rear. In V-type engines the numbering sequence varies with the manufacturer.

The firing order (which is different from the numbering order) of the cylinders is usually stamped on the cylinder block or on the manufacturer's nameplate.

VALVE ARRANGEMENT

The majority of internal combustion engines also are classified according to the position and arrangement of the intake and exhaust valves. This classification depends on whether the valves are in the cylinder block or in the cylinder head. Various arrangements have been used; the most common are the L-head, I-head, and F-head (fig. 12-8). The letter designation is used because the shape of the combustion chamber resembles the form of the letter identifying it.

L-Head

In the L-head engines, both valves are placed in the block on the same side of the cylinder. The valve-operating mechanism is located directly below the valves, and one camshaft actuates both the intake and exhaust valves.

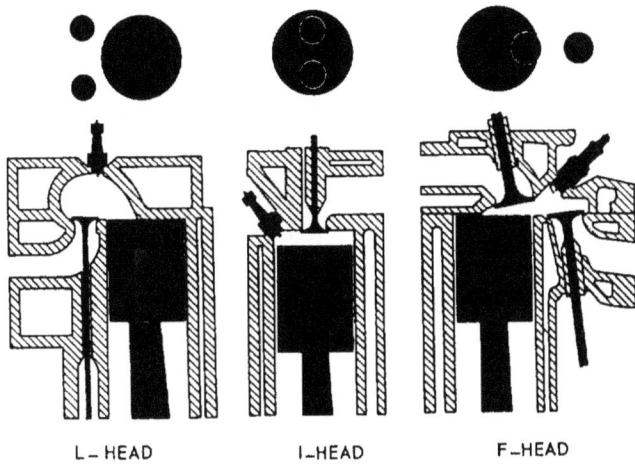

Figure 12-8.-L-, I-, and F-valve arrangement.

I-Head

Engines using the I-head construction are called valve-in-head or overhead valve engines, because the valves mount in a cylinder head above the cylinder. This arrangement requires a tappet, a push rod, and a rocker arm above the cylinder to reverse the direction of the valve movement. Only one camshaft is required for both valves. Some overhead valve engines make use of an overhead camshaft. This arrangement eliminates the long linkage between the camshaft and the valve.

F-Head

In the F-head engine, the intake valves normally are located in the head, while the exhaust valves are located in the engine block. This arrangement combines, in effect, the L-head and the I-head valve arrangements. The valves in the head are actuated from the camshaft through tappets, push rods, and rocker arms (I-head arrangement), while the valves in the block are actuated directly from the camshaft by tappets (L-head arrangement).

ENGINE CONSTRUCTION

Basic engine construction varies little, regardless of the size and design of the engine. The intended use of an engine must be considered before the design and size can be determined. The temperature at which an engine will operate has a great deal to do with the metals used in its construction.

The problem of obtaining servicing procedures and service parts in the field are simplified by the categorization of engines into families based on construction and design. Because many kinds of engines are needed for many different jobs, engines are designed to have closely related cylinder sizes, valve arrangements, and so forth. As an example, the General Motors series 71 engines may have two, three, four, or six cylinders. However, they are designed so that the same pistons, connecting rods, bearings, valves and valve operating mechanisms can be used in all four engines.

Engine construction, in this chapter, will be broken down into two categories: stationary parts and moving parts.

STATIONARY PARTS

The stationary parts of an engine include the cylinder block, cylinders, cylinder head or heads, crankcase, and the exhaust and intake manifolds. These parts furnish the framework of the engine. All movable parts are attached to or fitted into this framework.

Engine Cylinder Block

The engine cylinder block is the basic frame of a liquid-cooled engine, whether it is the in-line, horizontally opposed, or V-type. The cylinder block and crankcase are often cast in one piece that is the heaviest single piece of metal in the engine. (See fig. 12-9.) In small engines, where weight is an important consideration, the crankcase may be cast separately. In most large diesel engines, such as those used in power plants, the crankcase is cast separately and is attached to a heavy stationary engine base.

In practically all automotive and construction equipment, the cylinder block and crankcase are cast in one piece. In this course we are concerned primarily with liquid-cooled engines of this type.

The cylinders of a liquid-cooled engine are surrounded by jackets through which the cooling liquid circulates. These jackets are cast integrally with the cylinder block. Communicating passages permit the coolant to circulate around the cylinders and through the head.

The air-cooled engine cylinder differs from that of a liquid-cooled engine in that the cylinders are made individually, rather than cast in block. The cylinders of air-cooled engines have closely spaced fins surrounding the barrel; these fins provide an increased surface area from which heat can be dissipated. This engine design is in contrast to that of the liquid-cooled engine, which has a water jacket around its cylinders.

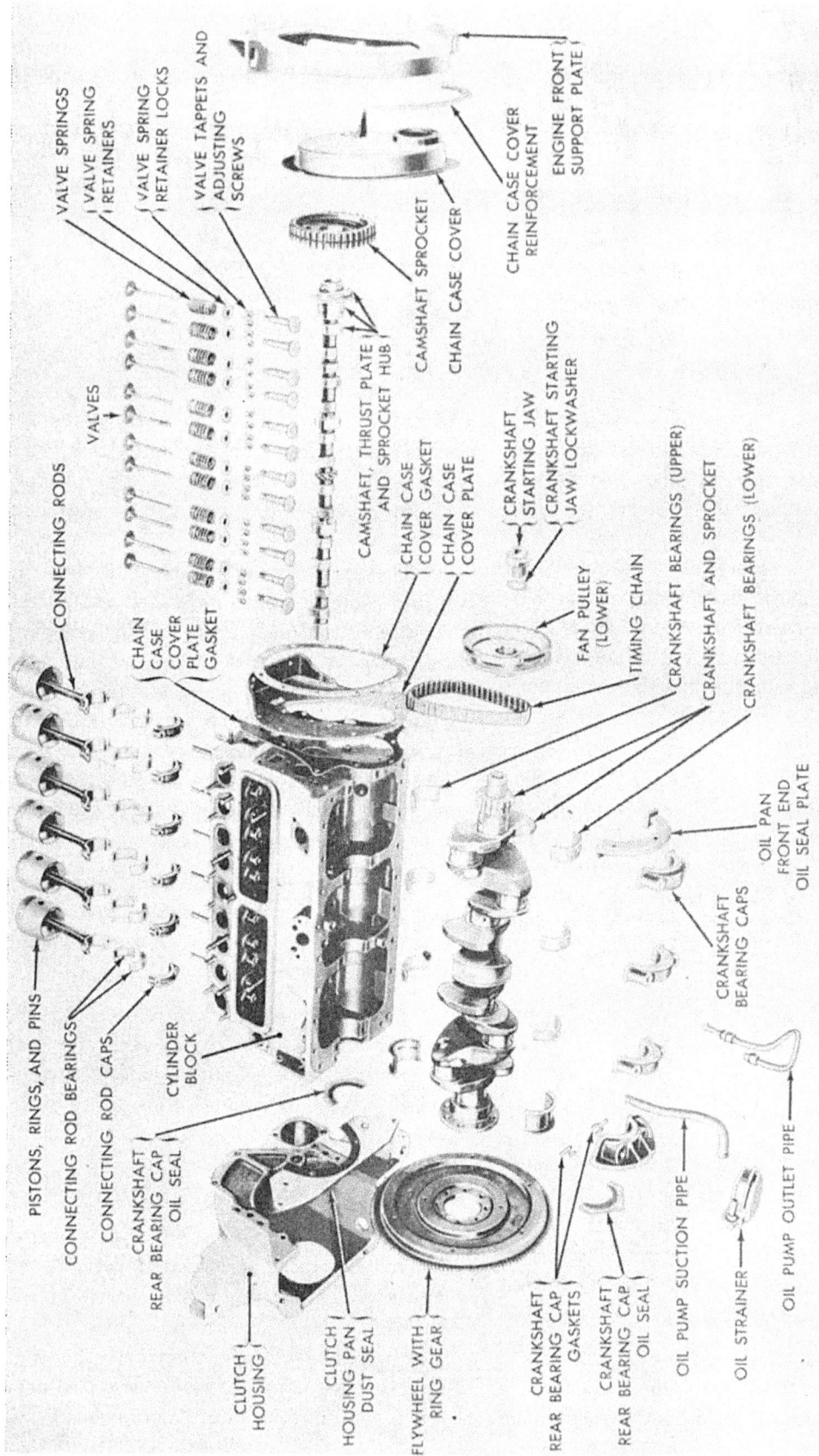

Figure 12-9.—Cylinder block and components.

VALVE SPRINGS
VALVE SPRING RETAINERS
VALVE SPRING RETAINER LOCKS
VALVE TAPPETS AND ADJUSTING SCREWS

CAMSHAFT SPROCKET

CHAIN CASE COVER

CHAIN CASE COVER REINFORCEMENT

ENGINE FRONT SUPPORT PLATE

VALVES

CONNECTING RODS

CHAIN CASE COVER PLATE GASKET

CAMSHAFT, THRUST PLATE AND SPROCKET HUB

CHAIN CASE COVER GASKET

CHAIN CASE COVER PLATE

CRANKSHAFT STARTING JAW

CRANKSHAFT STARTING JAW LOCKWASHER

FAN PULLEY (LOWER)

TIMING CHAIN

CRANKSHAFT BEARINGS (UPPER)

CRANKSHAFT AND SPROCKET

CRANKSHAFT BEARINGS (LOWER)

PISTONS, RINGS, AND PINS
CONNECTING ROD BEARINGS
CONNECTING ROD CAPS

CRANKSHAFT REAR BEARING CAP OIL SEAL

CYLINDER BLOCK

CLUTCH HOUSING

CLUTCH HOUSING PAN DUST SEAL

FLYWHEEL WITH RING GEAR

CRANKSHAFT REAR BEARING CAP GASKETS

CRANKSHAFT REAR BEARING CAP OIL SEAL

OIL PUMP SUCTION PIPE

OIL STRAINER

OIL PUMP OUTLET PIPE

CRANKSHAFT BEARING CAPS

OIL PAN FRONT END OIL SEAL PLATE

12-10

Cylinder Block Construction

The cylinder block is cast from gray iron or iron alloyed with other metals such as nickel, chromium, or molybdenum. Some lightweight engine blocks are made from aluminum.

Cylinders are machined by grinding or boring to give them the desired true inner surface. During normal engine operation, cylinder walls will wear out-of-round, or they may become cracked and scored if not properly lubricated or cooled. Liners (sleeves) made of metal alloys resistant to wear are used in many gasoline engines and practically all diesel engines to lessen wear. After they have been worn beyond the maximum oversize, the liners can be replaced individually, which permits the use of standard pistons and rings. Thus, you can avoid replacing the entire cylinder block

The liners are inserted into a hole in the block with either a PRESS FIT or a SLIP FIT. Liners are further designated as either a WET-TYPE or DRY-TYPE. The wet-type liner comes in direct contact with the coolant and is sealed at the top by a metallic sealing ring and at the bottom by a rubber sealing ring. The dry-type liner does not contact the coolant.

Engine blocks for L-head engines contain the passageways for the valves and valve ports. The lower part of the block (crankcase) supports the crankshaft (the main bearings and bearing caps) and provides a place to which the oil pan can be fastened.

The camshaft is supported in the cylinder block by bushings that fit into machined holes in the block. On L-head in-line engines, the intake and exhaust manifolds are attached to the side of the cylinder block. On L-head V-8 engines, the intake manifold is located between the two banks of cylinders; these engines have two exhaust manifolds, one on the outside of each bank.

Cylinder Head

The cylinder head provides the combustion chambers for the engine cylinders. It is built to conform to the arrangement of the valves: L-head, I-head, or other.

In the water-cooled engine, the cylinder head (fig. 12-10) is bolted to the top of the cylinder block to close the upper end of the cylinders. It contains passages,

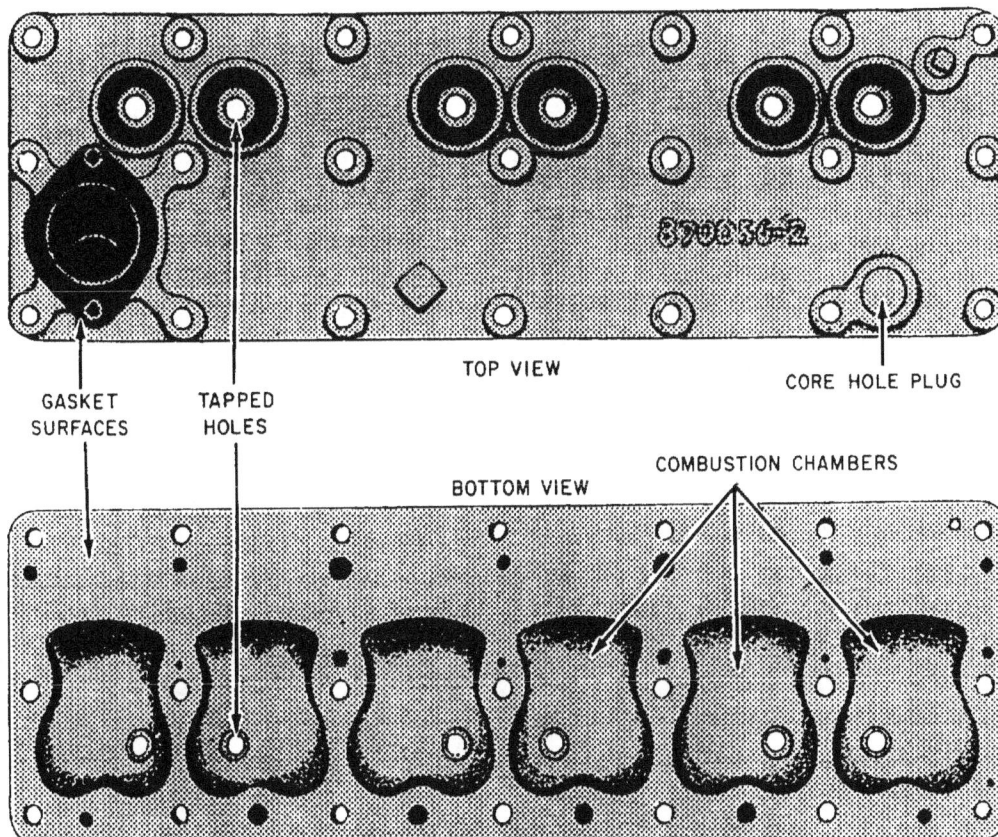

Figure 12-10-Cylinder head for L-head engine.

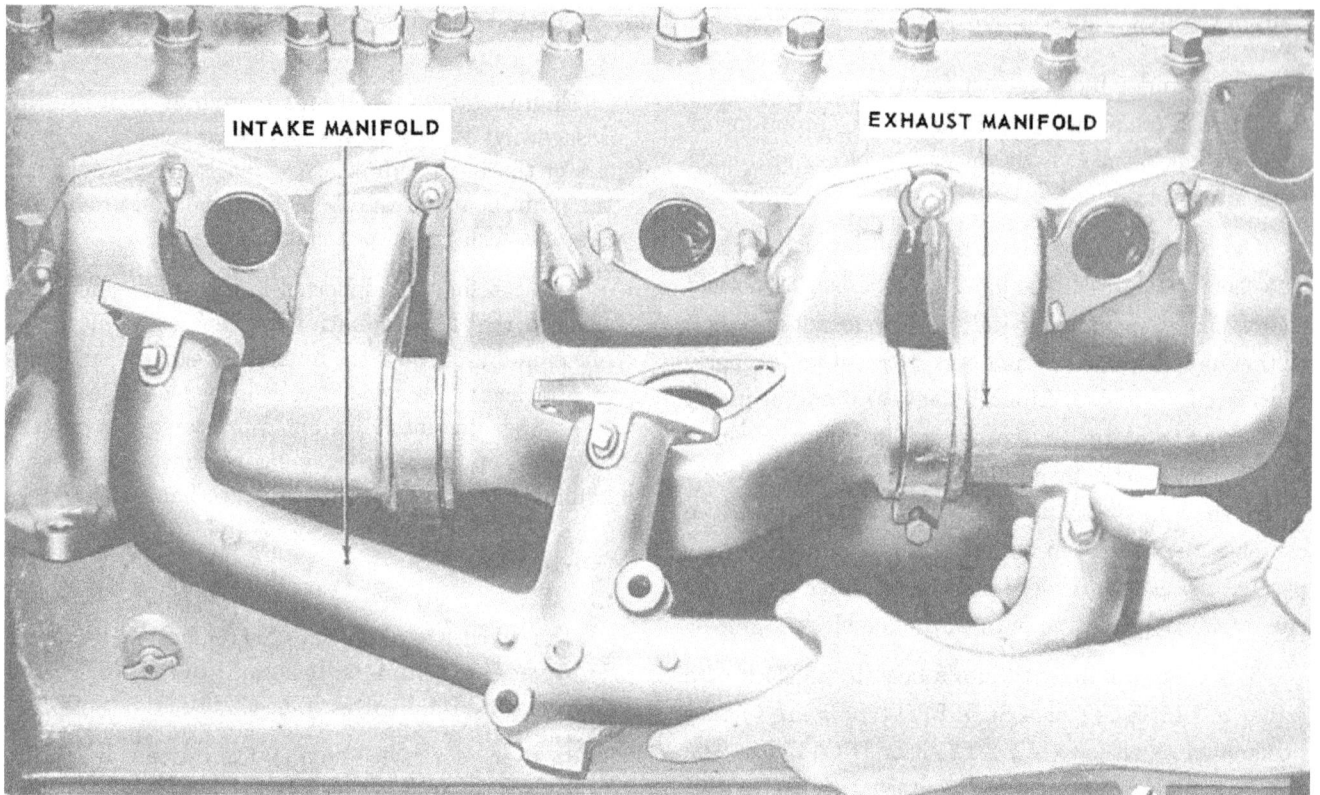

Figure 12-11.—Intake and exhaust manifolds.

matching those of the cylinder block, that allow the cooling water to circulate in the head. The head also helps keep compression in the cylinders. The gasoline engine contains tapped holes in the cylinder head that lead into the combustion chamber. The spark plugs are inserted into these tapped holes.

In the diesel engine the cylinder head may be cast in a single unit, or it may be cast for a single cylinder or two or more cylinders. Separated head sections (usually covering one, two, or three cylinders in large engines) are easy to handle and can be removed.

The L-head type of cylinder head shown in figure 12-10 is a comparatively simple casting. It contains water jackets for cooling, openings for spark plugs, and pockets into which the valves operate. Each pocket serves as a part of the combustion chamber. The fuel-air mixture is compressed in the pocket as the piston reaches the end of the compression stroke. Note that the pockets have a rather complex curved surface. This shape has been carefully designed so that the fuel-air mixture, compressed, will be subjected to violent turbulence. This turbulence ensures uniform mixing of the fuel and air, thus improving the combustion process.

The I-head (overhead-valve) type of cylinder head contains not only valve and combustion chamber pockets and water jackets for cooling spark-plug openings, but it also contains and supports the valves and valve-operating mechanisms. In this type of cylinder head, the water jackets must be large enough to cool not only the top of the combustion chamber but also the valve seats, valves, and valve-operating mechanisms.

Crankcase

The crankcase is that part of the engine block below the cylinders. It supports and encloses the crankshaft and provides a reservoir for the lubricating oil. Often times the crankcase contains a place for mounting the oil pump, oil filter, starting motor, and generator. The lower part of the crankcase is the OIL PAN, which is bolted at the bottom. The oil pan is made of pressed or cast steel and holds from 4 to 9 quarts of oil, depending on the engine design.

The crankcase also has mounting brackets that support the entire engine on the vehicle frame. These brackets are either an integral part of the crankcase or are bolted to it so that they support the engine at three or four points. These points of contact usually are cushioned with rubber that insulates the frame and the body of the vehicle from engine vibration and therefore prevents damage to the engine supports and the transmission.

Exhaust Manifold

The exhaust manifold is a tube that carries waste products of combustion from the cylinders. On L-head engines the exhaust manifold is bolted to the side of the engine block on; overhead-valve engines it is bolted to the side of the engine cylinder head. Exhaust manifolds may be single iron castings or may be cast in sections. They have a smooth interior surface with no abrupt change in size (see fig. 12-1 1).

Intake Manifold

The intake manifold on a gasoline engine carries the fuel-air mixture from the carburetor and distributes it as evenly as possible to the cylinders. On a diesel engine, the manifold carries only air to the cylinders. The intake manifold is attached to the block on L-head engines and to the side of the cylinder head on overhead-valve engines. (See fig. 12-11.)

In gasoline engines, smooth and efficient engine performance depends largely on whether the fuel-air mixtures that enter each cylinder are uniform in strength, quality, and degree of vaporization. The inside walls of the manifold must be smooth to offer little obstruction to the flow of the fuel-air mixture. The manifold is designed to prevent the collecting of fuel at the bends in the manifold.

The intake manifold should be as short and straight as possible to reduce the chances of condensation between the carburetor and cylinders. Some intake manifolds are designed so that hot exhaust gases heat their surfaces to help vaporize the fuel.

Gaskets

The principal stationary parts of an engine have just been explained. The gaskets (fig. 12- 12) that serve as seals between these parts require as much attention during engine assembly as any other part. It is impractical to machine all surfaces so that they fit together to form a perfect seal. The gaskets make a joint that is air, water, or oil tight. Therefore, when properly

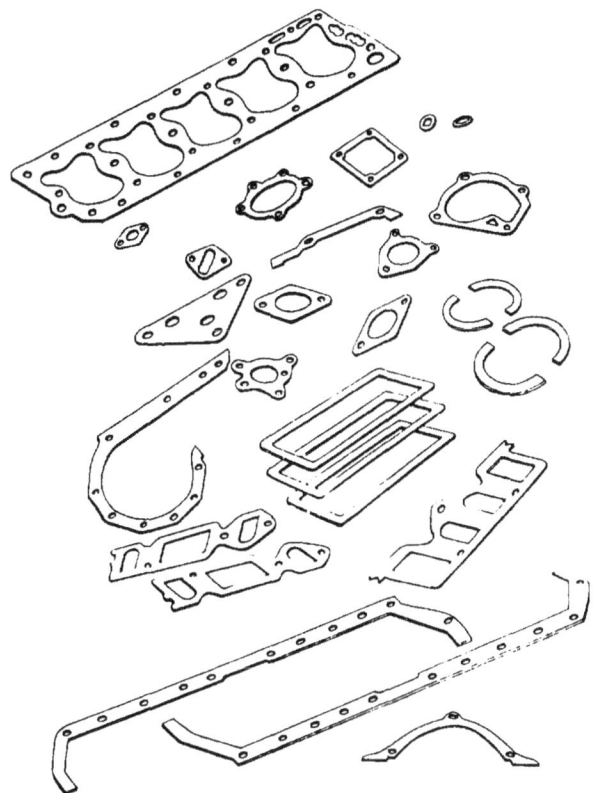

Figure 12-12.-Engine overhaul gasket kit.

installed, they prevent loss of compression, coolant, or lubricant.

MOVING PARTS OF AN ENGINE

The moving parts of an engine serve an important function in turning heat energy into mechanical energy. They further convert reciprocal motion into rotary motion. The principal moving parts are the piston assembly, connecting rods, crankshaft assembly (includes flywheel and vibration dampener), camshaft, valves, and gear train.

The burning of the fuel-air mixture within the cylinder exerts a pressure on the piston, thus pushing it down in the cylinder. The action of the connecting rod and crankshaft converts this downward motion to a rotary motion.

Piston Assembly

Engine pistons serve several purposes. They transmit the force of combustion to the crankshaft through the connecting rod. They act as a guide for the upper end of the connecting rod. And they also serve as

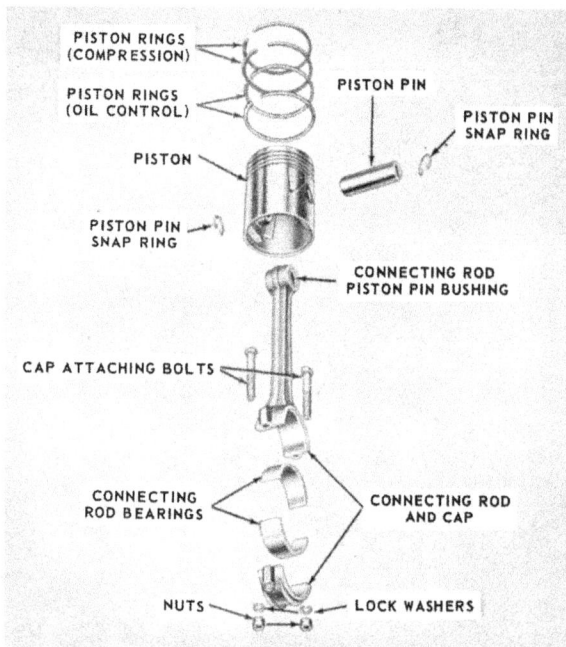

Figure 12-13.—Piston and connecting rod (exploded view).

a carrier for the piston rings used to seal the compression in the cylinder. (See. fig. 12-13.)

The piston must come to a complete stop at the end of each stroke before reversing its course in the cylinder. To withstand this rugged treatment and wear, it must be made of tough material, yet be light in weight. To overcome inertia and momentum at high speed, it must be carefully balanced and weighed. All the pistons used in any one engine must be of similar weight to avoid excessive vibration. Ribs are used on the underside of the piston to reinforce the hand. The ribs also help to conduct heat from the head of the piston to the piston rings and out through the cylinder walls.

The structural components of the piston are the head, skirt, ring grooves, and land (fig. 12-14). However, all pistons do not look like the typical one illustrated here. Some have differently shaped heads. Diesel engine pistons usually have more ring grooves and rings than gasoline engine pistons. Some of these rings may be installed below as well as above the wrist or piston pin (fig. 12-15).

Fitting pistons properly is important. Because metal expands when heated and space must be provided for lubricants between the pistons and the cylinder walls, the pistons are fitted to the engine with a specified clearance. This clearance depends upon the size or diameter of the piston and the material form which it is made. Cast iron does not expand as fast or as much as aluminum. Aluminum pistons require more clearance to prevent binding or seizing when the engine gets hot. The skirt of bottom part of the piston runs much cooler than the top; therefore, it does not require as much clearance as the head.

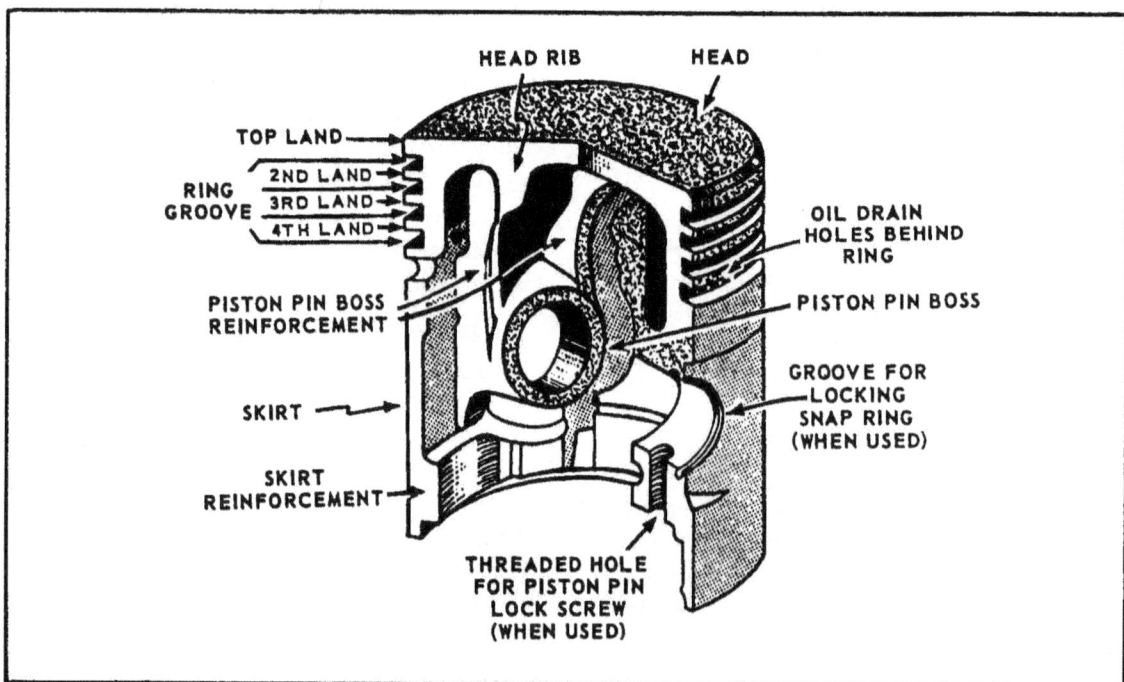

Figure 12-14.—The parts of a piston.

12-14

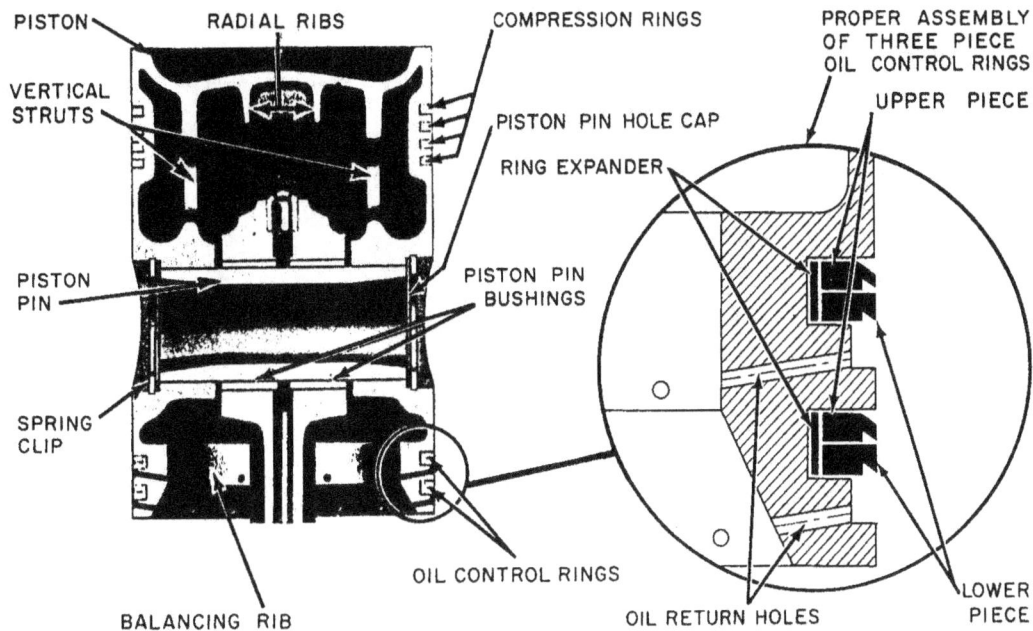

Figure 12-15.—Piston assembly.

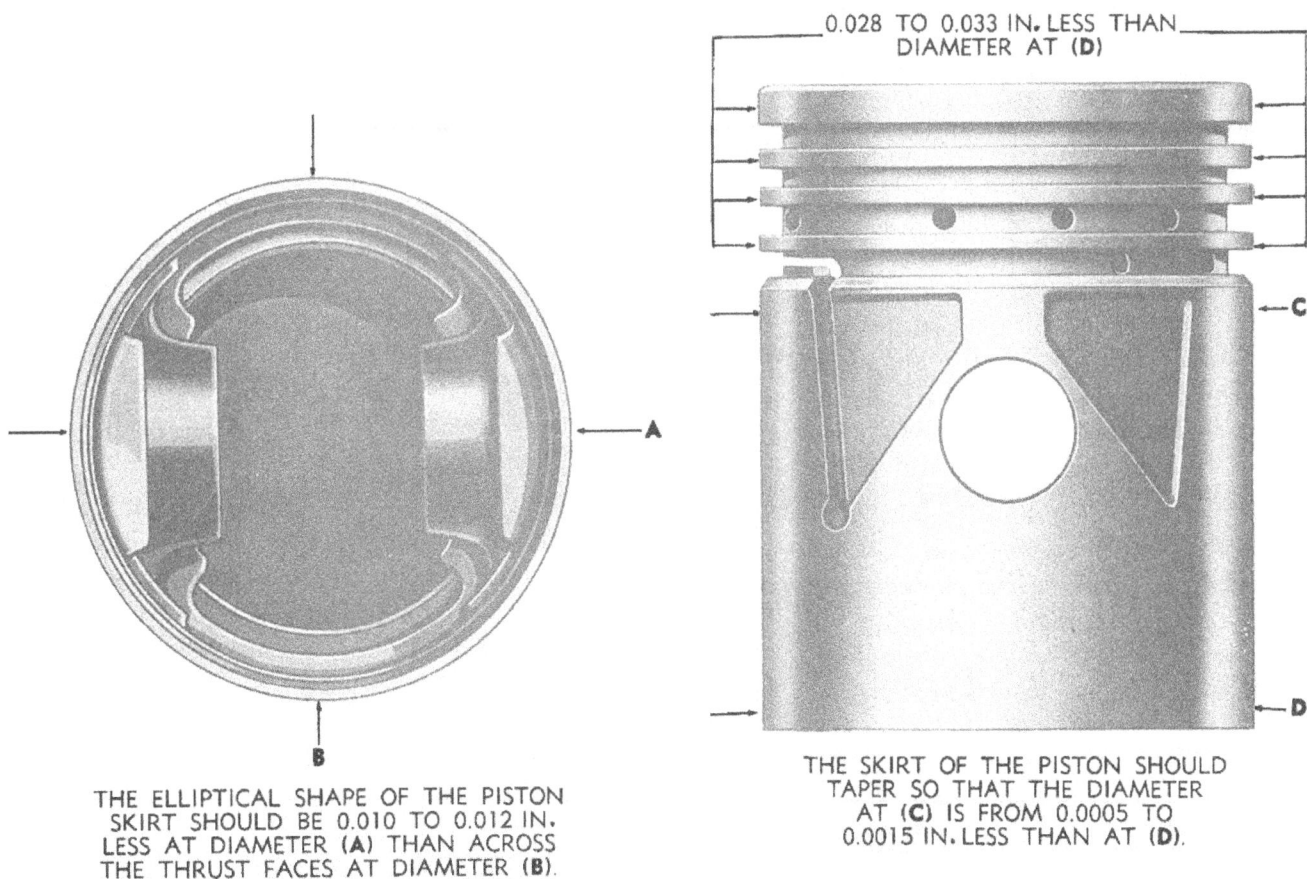

THE ELLIPTICAL SHAPE OF THE PISTON
SKIRT SHOULD BE 0.010 TO 0.012 IN.
LESS AT DIAMETER (A) THAN ACROSS
THE THRUST FACES AT DIAMETER (B).

THE SKIRT OF THE PISTON SHOULD
TAPER SO THAT THE DIAMETER
AT (C) IS FROM 0.0005 TO
0.0015 IN. LESS THAN AT (D).

Figure 12-16.—Cam-ground piston.

The piston is kept in alignment by the skirt, which is usually cam ground (elliptical in cross section) (fig.12-16). This elliptical shape permits the piston to fit the cylinder, regardless of whether the piston is cold or at operating temperature. The narrowest diameter of the piston is at the piston pin bosses, where the piston skirt is thickest. At the widest diameter of the piston, the piston skirt is thinnest. The piston is fitted to close limits at its widest diameter so that the piston noise (slap) is prevented during the engine warm-up. As the piston is

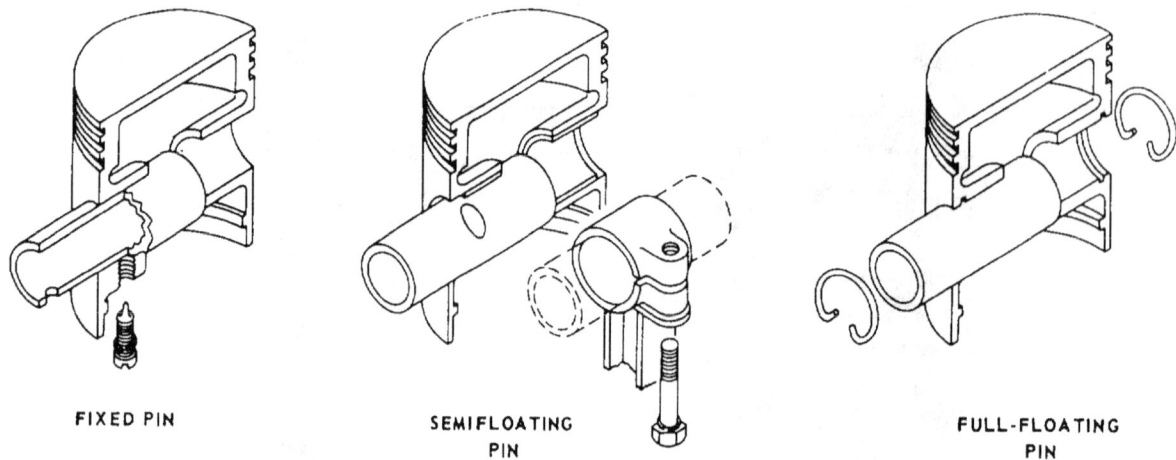

Figure 12-17.-Piston pin types.

expanded by the heat generated during operation, it becomes round because the expansion is proportional to the temperature of the metal. The walls of the skirt are cut away as much as possible to reduce weight and to prevent excessive expansion during engine operation. Many aluminum pistons are made with split skirts so that when the pistons expand, the skirt diameter will not increase.

The two types of piston skirts found in most engines are the full trunk and the slipper. The full-trunk-type skirt, more widely used, has a full cylindrical shape with bearing surfaces parallel to those of the cylinder, giving more strength and better control of the oil film. The slipper-type (cutaway) skirt has considerable relief on the sides of the skirt, leaving less area for possible contact with the cylinder walls and thereby reducing friction.

PISTON PINS.— The piston is attached to the connecting rod by the piston pin (wrist pin). The pin passes through the piston pin bosses and through the upper end of the connecting rod, which rides within the piston on the middle of the pin. Piston pins are made of alloy steel with a precision finish and are case hardened and sometimes chromium plated to increase their wearing qualities. Their tubular construction gives them maximum strength with minimum weight. They are lubricated by splash from the crankcase or by pressure through passages bored in the connecting rods.

Three methods are commonly used for fastening a piston pin to the piston and the connecting rod: fixed pin, semifloating pin, and full-floating pin (fig. 12-17). The anchored, or fixed, pin attaches to the piston by a screw running through one of the bosses; the connecting rod oscillates on the pin. The semifloating pin is

anchored to the connecting rod and turns in the piston pin bosses. The full-floating pin is free to rotate in the connecting rod and in the bosses, while plugs or snap-ring locks prevent it from working out against the sides of the cylinder.

PISTON RINGS.— Piston rings are used on pistons to maintain gastight seals between the pistons and cylinders, to aid in cooling the piston, and to control cylinder-wall lubrication. About one-third of the heat absorbed by the piston passes through the rings to the cylinder wall. Piston rings are often complicated in design, are heat treated in various ways, and are plated with other metals. Piston rings are of two distinct classifications: compression rings and oil control rings. (See fig. 12-18.)

The principal function of a compression ring is to prevent gases from leaking by the piston during the compression and power strokes. All piston rings are split to permit assembly to the piston and to allow for expansion. When the ring is in place, the ends of the split joint do not form a perfect seal; therefore, more than one ring must be used, and the joints must be staggered around the piston. If cylinders are worn, expanders (figs. 12-15 and 12-18) are sometimes used to ensure a perfect seal.

The bottom ring, usually located just above the piston pin, is an oil-regulating ring. This ring scrapes the excess oil from the cylinder walls and returns some of it, through slots, to the piston ring grooves. The ring groove under an oil ring has openings through which the oil flows back into the crankcase. In some engines, additional oil rings are used in the piston skirt below the piston pin.

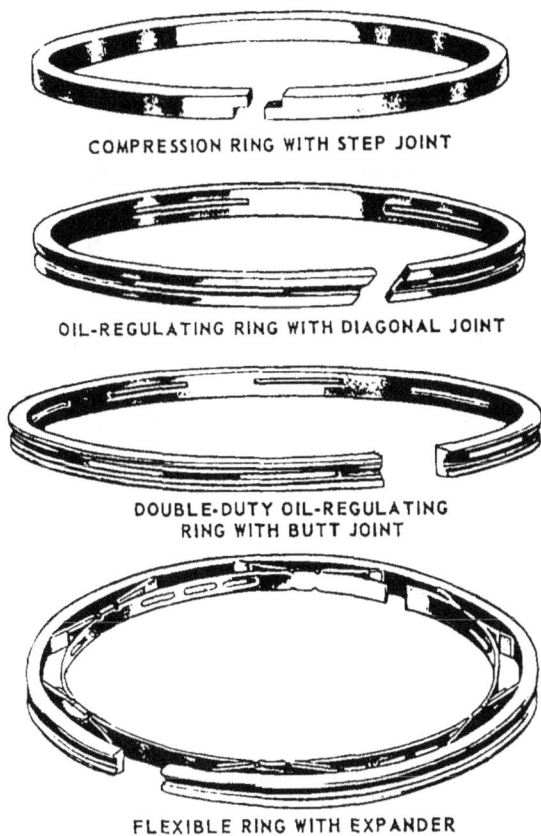

COMPRESSION RING WITH STEP JOINT

OIL-REGULATING RING WITH DIAGONAL JOINT

DOUBLE-DUTY OIL-REGULATING
RING WITH BUTT JOINT

FLEXIBLE RING WITH EXPANDER

Figure 12-18.-Piston rings.

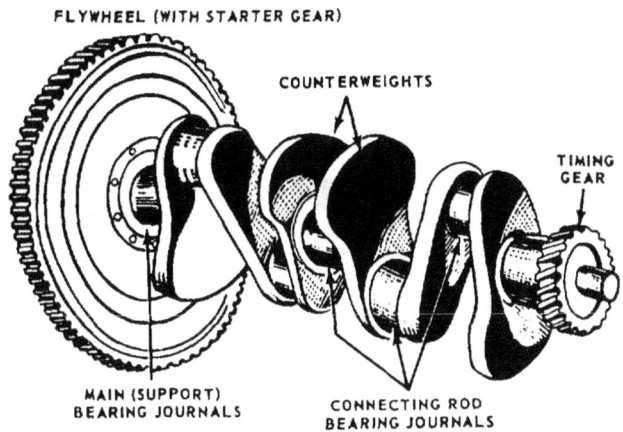

Figure 12-19.-Crankshaft of a four-cylinder engine.

Connecting Rods

Connecting rods must be light and yet strong enough to transmit the thrust of the pistons to the crankshaft. Connecting rods are drop forged from a steel alloy capable of withstanding heavy loads without bending or twisting. Holes at the upper and lower ends are machined to permit accurate fitting of bearings. These holes must be parallel.

The upper end of the connecting rod is connected to the piston by the piston pin. If the piston pin is locked in the piston pin bosses or if it floats in both the piston and the connecting rod, the upper hold of the connecting rod will have a solid bearing (bushing) of bronze or similar material. As the lower end of the connecting rod revolves with the crankshaft, the upper end is forced to turn back and forth on the piston pin. Although this movement is slight, the bushing is necessary because of the high pressure and temperatures. If the piston pin is semifloating, a bushing is not needed.

The lower hole in the connecting rod is split to permit it to be clamped around the crankshaft. The bottom part, or cap, is made of the same material as the rod and is attached by two or more bolts. The surface that bears on the crankshaft is generally a bearing material in the form of a separate split shell; in a few cases, it may be spun or die-cast in the inside of the rod and cap during manufacture. The two parts of the separate bearing are positioned in the rod and cap by dowel pins, projections, or short brass screws. Split bearings may be of the precision or semiprecision type.

The precision type bearing is accurately finished to fit the crankpin and does not require further fitting during installation. It is positioned by projections on the shell that match reliefs in the rod and cap. The projections prevent the bearings from moving sideways and prevent rotary motion in the rod and cap.

The semiprecision-type bearing is usually fastened to or die-cast with the rod and cap. Before installation, it is machined and fitted to the proper inside diameter with the cap and rod bolted together.

Crankshaft

As the pistons collectively might be regarded as the heart of the engine, so the crankshaft might be considered the backbone (fig. 12-19). It ties together the reactions of the pistons and the connecting rods, transforming their reciprocating motion into rotary motion. It transmits engine power through the flywheel, clutch, transmission, and differential to drive your vehicle.

The crankshaft is forged or cast from an alloy of steel and nickel. It is machined smooth to provide

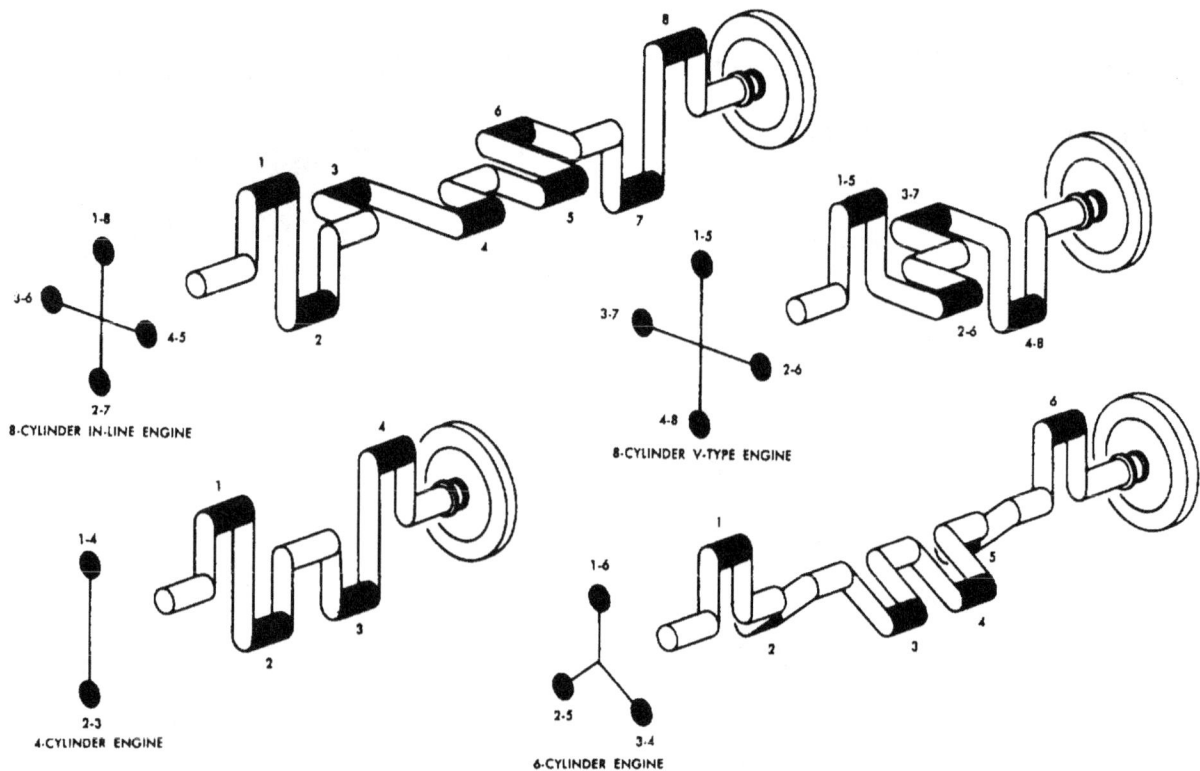

Figure 12-20.-Crankshaft and throw arrangements commonly used.

bearing surfaces for the connecting rods and the main bearings. It is case-hardened (coated in a furnace with copper alloyed and carbon). These bearing surfaces are called journals. The crankshaft counterweights impede the centrifugal force of the connecting rod and assembly attached to the throws or points of bearing support. These throws must be placed so that they counter-balance each other.

Crankshaft and throw arrangements for four-, six-, and eight-cylinder engines are shown in figure 12-20. Four-cylinder engine crankshafts have either three or five main support bearings and four throws in one plane. As shown in the figure, the four throws for the number 1 and 4 cylinders (four-cylinder engine) are 180° from those for the number 2 and 3 cylinders. On six-cylinder engine crankshafts, each of the three pairs of throws is arranged 120° from the other two. Such crankshafts may be supported by as many as seven main bearings—one at each end of the shaft and one between each pair of crankshaft throws. The crankshafts of eight-cylinder V-type engines are similar to those of the four-cylinder in-line type. They may have each of the four throws fixed at 90° from each other (as in fig. 12-20) for better balance and smoother operation.

V-type engines usually have two connecting rods fastened side by side on one crankshaft throw. With this arrangement, one bank of the engine cylinders is set slightly ahead of the other to allow the two rods to clear each other.

Vibration Damper

The power impulses of an engine result in torsional vibration in the crankshaft. A vibration damper mounted on the front of the crankshaft controls this vibration (fig. 12-21). If this torsional vibration were not controlled, the crankshaft might actually break at certain speeds.

Most types of vibration dampers resemble a miniature clutch. A friction facing is mounted between the hub face and a small damper flywheel. The damper flywheel is mounted on the hub face with bolts that go through rubber cones in the flywheel. These cones permit limited circumferential movement between the crankshaft and damper flywheel. That reduces the effects of the torsional vibration in the crankshaft. Several other types of vibration dampers are used; however, they all operate in essentially the same way.

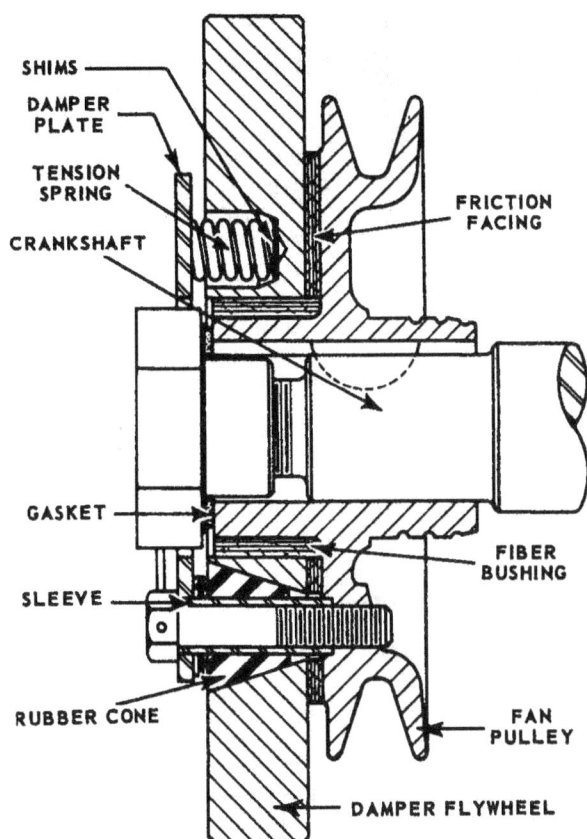

Figure 12-21.-Sectional view of a typical vibration damper.

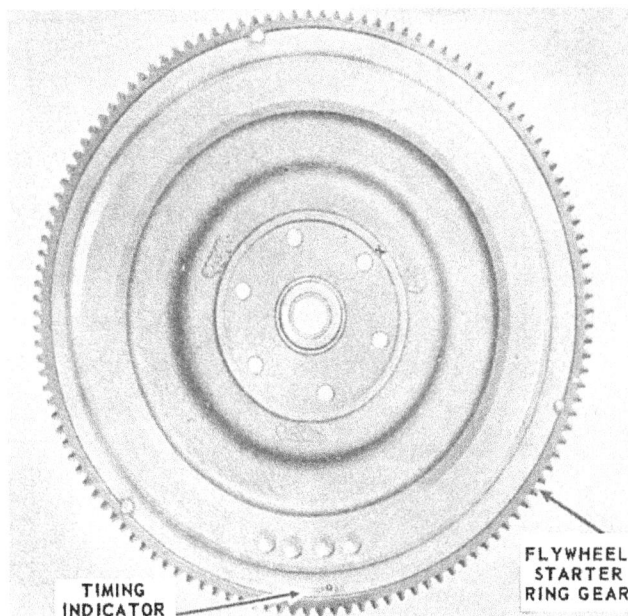

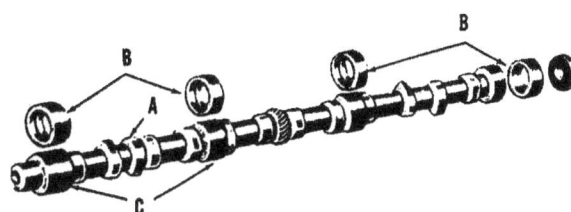

A – CAMSHAFT
B – CAMSHAFT BEARING
C – BEARING JOURNAL

Figure 12-23.-Camshaft and bushings.

Engine Flywheel

The flywheel mounts at the rear of the crankshaft near the rear main bearing. This is usually the longest and heaviest main bearing in the engine, as it must support the weight of the flywheel.

The flywheel (fig. 12-22) stores up rotation energy during the power impulses of the engine. It releases this energy between power impulses, thus assuring less fluctuation in engine speed and smoother engine operation. The size of the flywheel will vary with the number of cylinders and the general construction of the engine. With the large number of cylinders and the consequent overlapping of power impulses, there is less need for a flywheel; consequently, the flywheel can be relatively small. The flywheel rim carries a ring gear, either integral with or shrunk on the flywheel, that meshes with the starter driving gear for cranking the engine. The rear face of the flywheel is usually machined and ground and acts as one of the pressure surfaces for the clutch, becoming a part of the clutch assembly.

Valves and Valve Mechanisms

Most engines have two valves for each cylinder, one intake and one exhaust valve. Since each of these valves operates at different times, separate operating mechanisms must be provided for each valve. Valves are normally held closed by heavy springs and by compression in the combustion chamber. The purpose of the valve-actuating mechanism is to overcome the spring pressure and open the valves at the proper time. The valve-actuating mechanism includes the engine camshaft, camshaft followers (tappets), pushrods, and rocker arms.

CAMSHAFT.—The camshaft (fig. 12-23) is enclosed in the engine block. It has eccentric lobes (cams) ground on it for each valve in the engine. As the

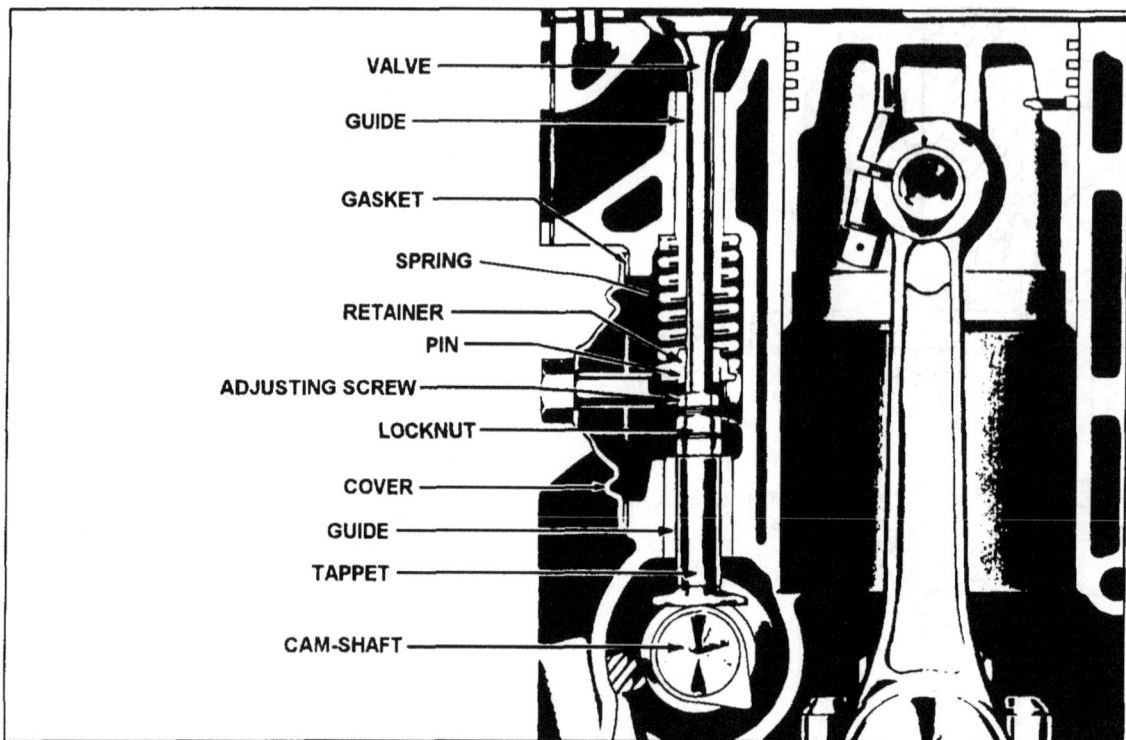

Figure 12-24.-L-head valve operating mechanism.

VALVE
GUIDE
GASKET
SPRING
RETAINER
PIN
ADJUSTING SCREW
LOCKNUT
COVER
GUIDE
TAPPET
CAM-SHAFT

camshaft rotates, the cam lobe moves up under the valve tappet, exerting an upward thrust through the tappet against the valve stem or a pushrod. This thrust overcomes the valve spring pressure as well as the gas pressure in the cylinder, causing the valve to open. When the lobe moves from under the tappet, the valve spring pressure reseats the valve.

On L-, F-, or I-head engines, the camshaft is usually located to one side and above the crankshaft; in V-type engines, it is usually located directly above the crankshaft. On the overhead camshaft engine, such as the Murphy diesel, the camshaft is located above the cylinder head.

The camshaft of a four-stroke cycle engine turns at one-half engine speed. It is driven off the crankshaft through timing gears or a timing chain. In the two-stroke cycle engine, the camshaft must turn at the same speed as the crankshaft so that each valve may open and close once in each revolution of the engine.

In most cases the camshaft will do more than operate the valve mechanism. It may have extra cams or gears that operate fuel pumps, fuel injectors, the ignition distributor, or the lubrication pump.

Camshafts are supported in the engine block by journals in bearings. Camshaft bearing journals are the hugest machined surfaces on the shaft. The bearings are usually made of bronze and are bushings rather than split bearings. The bushings are lubricated by oil circulating through drilled passages from the crankcase. The stresses on the camshaft are small; therefore, the bushings are not adjustable and require little attention. The camshaft bushings are replaced only when the engine requires a complete overhaul.

FOLLOWERS.— Camshaft followers are the parts of the valve-actuating mechanism (figs. 12-24 and 12-25) that contact the camshaft. You will probably hear them called valve tappets or vale lifters. In the L-head engine, the followers directly contact the end of the valve stem and have an adjusting device in them. In the overhead valve engine, the followers contact the pushrod that operates the rocker arm. The end of the rocker arm opposite the pushrod contacts the valve stem. The valve adjusting device, in this case, is in the rocker arm.

Many engines have self-adjusting, hydraulic valve lifters that always operate at zero clearance.

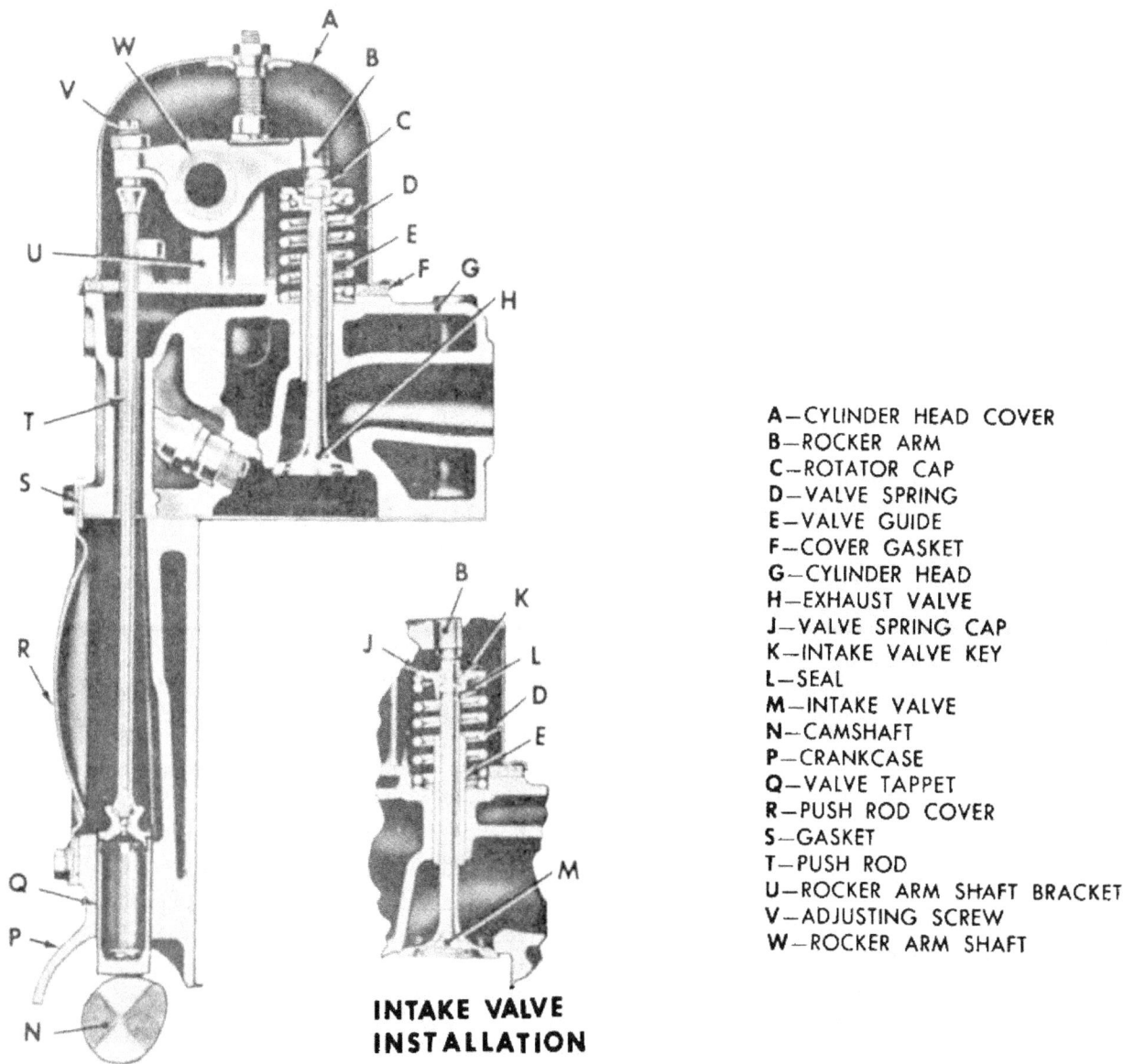

A—CYLINDER HEAD COVER
B—ROCKER ARM
C—ROTATOR CAP
D—VALVE SPRING
E—VALVE GUIDE
F—COVER GASKET
G—CYLINDER HEAD
H—EXHAUST VALVE
J—VALVE SPRING CAP
K—INTAKE VALVE KEY
L—SEAL
M—INTAKE VALVE
N—CAMSHAFT
P—CRANKCASE
Q—VALVE TAPPET
R—PUSH ROD COVER
S—GASKET
T—PUSH ROD
U—ROCKER ARM SHAFT BRACKET
V—ADJUSTING SCREW
W—ROCKER ARM SHAFT

INTAKE VALVE
INSTALLATION

Figure 12-25.—Valve operating mechanism for an overhead valve engine.

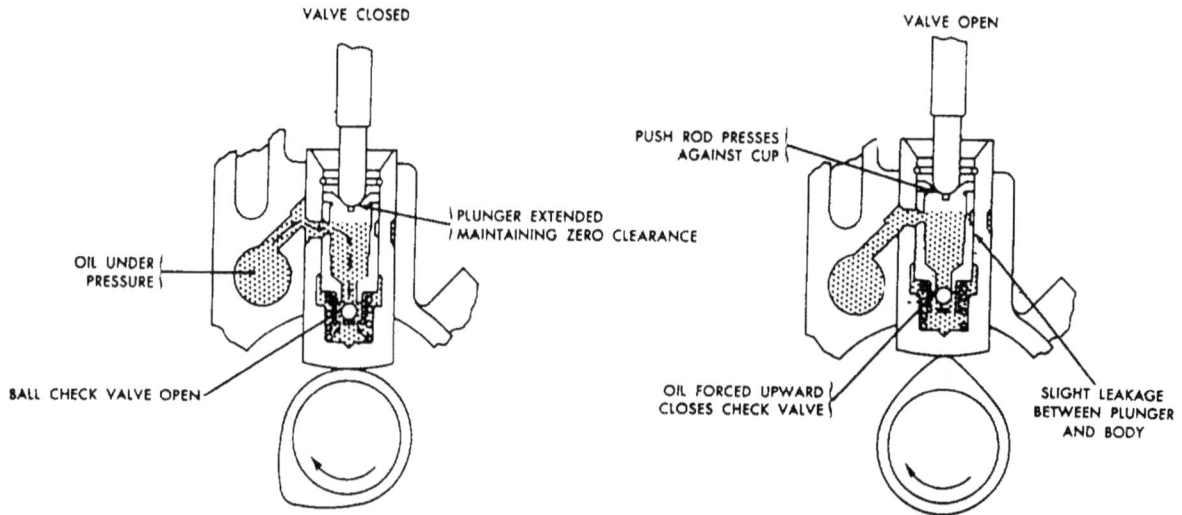

Figure 12-26.-Operation of a hydraulic valve lifter.

Figure 12-26 shows the operation of one type of hydraulic valve tappet mechanism. Oil under pressure is forced into the tappet when the valve is closed. This pressure extends the plunger in the tappet so that all valve clearance, or lash, is eliminated. When the cam lobe moves around under the tappet and starts to raise it, you hear no tappet noise. The movement of the tappet forces the oil upward in the lower chamber of the tappet. This action closes the ball check valve so that oil cannot escape. Then the tappet acts as though it were a simple, one-piece tappet and the valve is opened. When the lobe moves out from under the tappet and the valve closes, the pressure in the lower chamber of the tappet is relieved. Any slight loss of oil from the lower chamber is replaced by the oil pressure from the engine lubricating system. This oil pressure causes the plunger to move up snugly against the push rod so that any clearance is eliminated.

Timing Gears (Gear Trains)

Timing gears keep the crankshaft and camshaft turning in proper relation to one another so that the valves open and close at the proper time. Some engines use sprockets and chains.

The gears or sprockets, as the case may be, of the camshaft and crankshaft are keyed into position so that they cannot slip. Since they are keyed to their respective shafts, they can be replaced if they become worn or noisy.

With directly driven timing gears (fig. 12-27), one gear usually has a mark on two adjacent teeth and the other a mark on only one tooth. Timing the valves properly requires that the gears mesh so that the two marked teeth of one gear straddle the single marked tooth of the other.

AUXILIARY ASSEMBLIES

We have discussed the main parts of the engine proper; but other parts, both moving and stationary, are essential to engine operation. They are not built into the engine itself, but usually are attached to the engine block or cylinder head.

The fuel system includes a fuel pump and carburetor mounted on the engine. In diesel engines the fuel injection mechanism replaces the carburetor. An

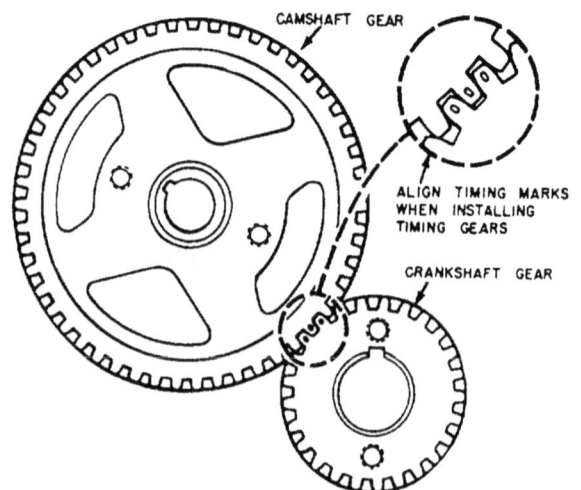

Figure 12-27.-Timing gears and their markings.

electrical system is provided to supply power for starting the engine and for igniting it during operation. The operation of an internal combustion engine requires an efficient cooling system. Water-cooled engines use a water pump and fan while air-cooled engines use a blower to force cool air around the engine cylinders.

In addition, an exhaust system is provided to carry away the burned gases exhausted from the engine cylinders. These systems will not be discussed in this course, however. For further information, refer to NAVPERS 10644G-1, *Construction Mechanic 3 & 2*.

SUMMARY

This chapter explained briefly the following operational principles and basic mechanisms of the internal combustion engine:

The power of an internal combustion engine comes from the burning of a mixture of fuel and air in a small, enclosed space.

The movement of the piston from top to bottom is called a stroke.

To produce sustained power, an engine must repeatedly accomplish a definite series of operations. This series of events is called a cycle.

Engine classifications are based on the type of fuel used—gasoline or diesel.

Design and size must be considered before engine construction.

Engines require the use of auxiliary assemblies such as the fuel pump, the carburetor, and an electrical system.

CHAPTER 13

POWER TRAINS

CHAPTER LEARNING OBJECTIVES

Upon completion of this chapter, you should be able to do the following:

● *Explain the mechanism of a power train.*

In chapter 12 we saw how a combination of simple machines and basic mechanisms was used in constructing the internal combustion engine. In this chapter we will learn how the power developed by the engine is transmitted to perform the work required of it. We will demonstrate the power train system used in automobiles and most trucks in our discussion. As we study the application of power trains, again look for the simple machines that make up each of the machines or mechanisms.

AUTOMOTIVE POWER TRAINS

In a vehicle, the mechanism that transmits the power of the engine to the wheels or tracks and accessory equipment is called the power train. In a simple situation, a set of gears or a chain and sprocket could perform this task, but automotive and construction vehicles are not usually designed for such simple operating conditions. They are designed to have great pulling power, to move at high speeds, to travel in reverse as well as forward, and to operate on rough terrain as well as smooth roads. To meet these widely varying demands, vehicles require several additional accessory units.

The power trains of automobiles and light trucks driven by the two rear wheels consist of a clutch, a transmission, a propeller shaft, a differential, and driving axles (fig. 13-1).

Four- and six-wheel drive trucks have transfer cases with additional drive shafts and live axles. Tractors, shovels, cranes, and other heavy-duty vehicles that move on tracks also have similar power trains. In addition to assemblies that drive sprockets to move the tracks, these vehicles also have auxiliary transmissions

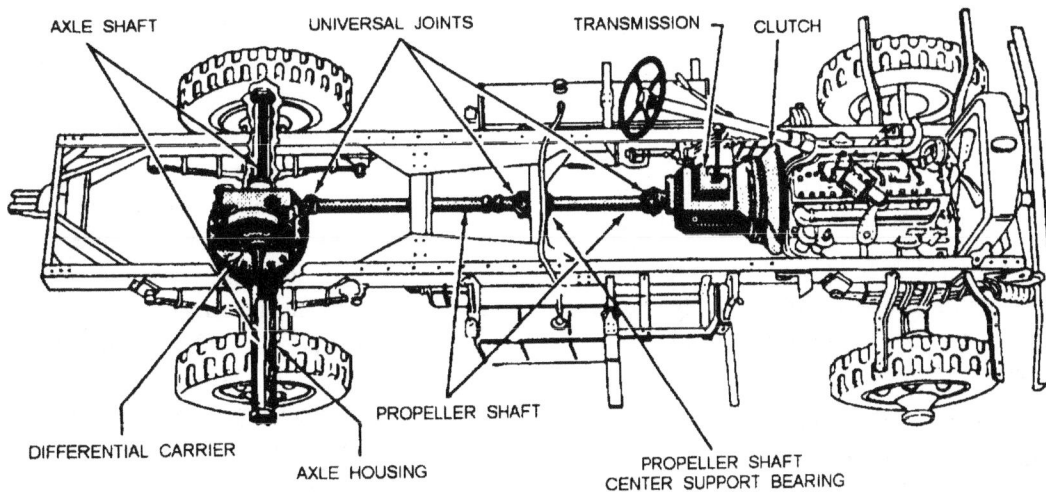

Figure 13-1.-Type of power transmission.

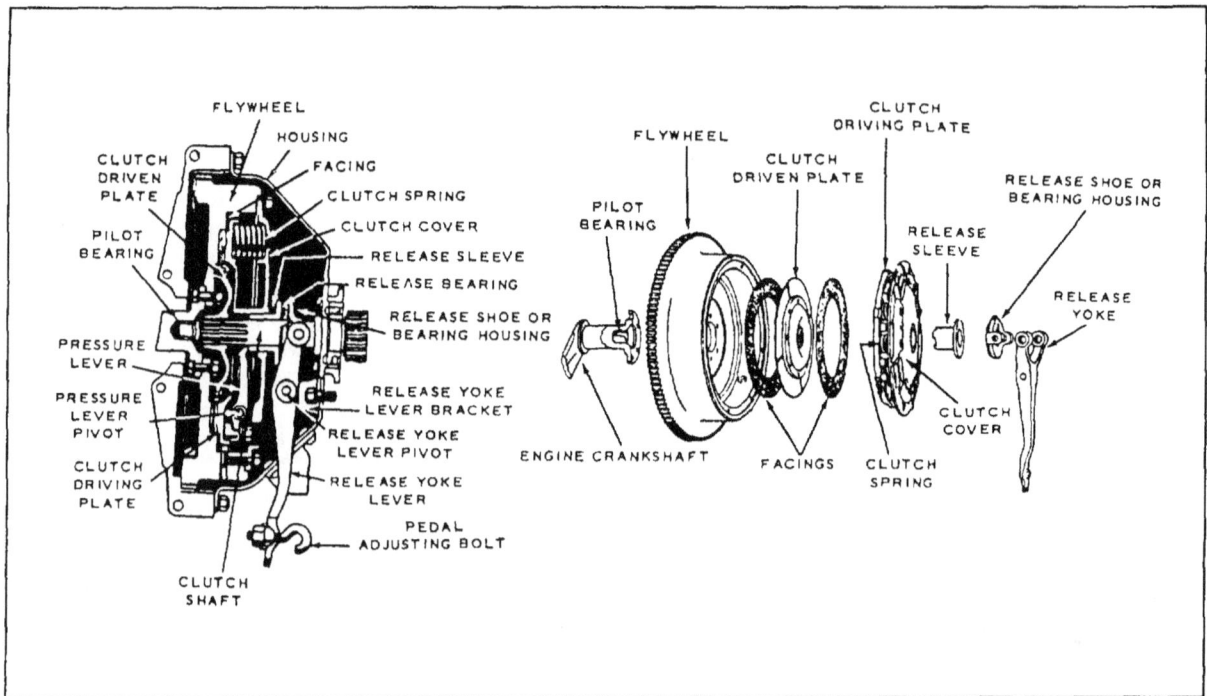

Figure 13-2.-Exploded and cross-sectional views of a plate clutch.

or power takeoff units. These units may be used to operate accessory attachments. The propeller shafts and clutch assemblies of these power trains are very much like those used to drive the wheels.

THE CLUTCH

The clutch is placed in the power train of motorized equipment for two purposes:

First, it provides a means of disconnecting the power of the engine from the driving wheels and accessory equipment. When you disengage the clutch, the engine can run without driving the vehicle or operating the accessories.

Second, when you start the vehicle, the clutch allows the engine to take up the load of driving the vehicle or accessories gradually and without shock.

Clutches are located in the power train between the source of power and the operating unit. Usually, they are placed between the engine and the transmission assembly, as shown in figure 13-1.

Clutches generally transmit power from the clutch-driving member to the driven member by friction. Strong springs within the plate clutch (fig. 13-2) gradually bring the driving member (plate), secured to the engine flywheel, in contact with the driven member

(disc). The driver of the automobile controls the pressure of the springs through use of the clutch. If the driver only applies light pressure, little friction takes place between the two members, which permits the clutch to slip. As the driver increases pressure, friction also increases and less slippage occurs. When the driver's foot releases pressure from the clutch pedal and applies full spring pressure, the driving plate and driven disc move at the same speed. All slipping then stops because of the direct connection between the driving and driven shafts.

In most clutches, a direct mechanical linkage exists between the clutch pedal and the clutch release yoke lever. Many late model vehicles and some larger units that require greater pressure to release the spring use a hydraulic clutch release system. A master cylinder (fig. 13-3), similar to the brake master cylinder, attaches to the clutch pedal. A cylinder, similar to a single-acting brake wheel cylinder, connects to the master cylinder by flexible pressure hose or metal tubing (fig. 13-3). The slave cylinder connects to the clutch release yoke lever. Movement of the clutch pedal actuates the clutch master cylinder. Hydraulic pressure transfers this movement to the slave cylinder, which, in turn, actuates the clutch release yoke lever.

We use various types of clutches. Most passenger cars and light trucks use the previously mentioned plate

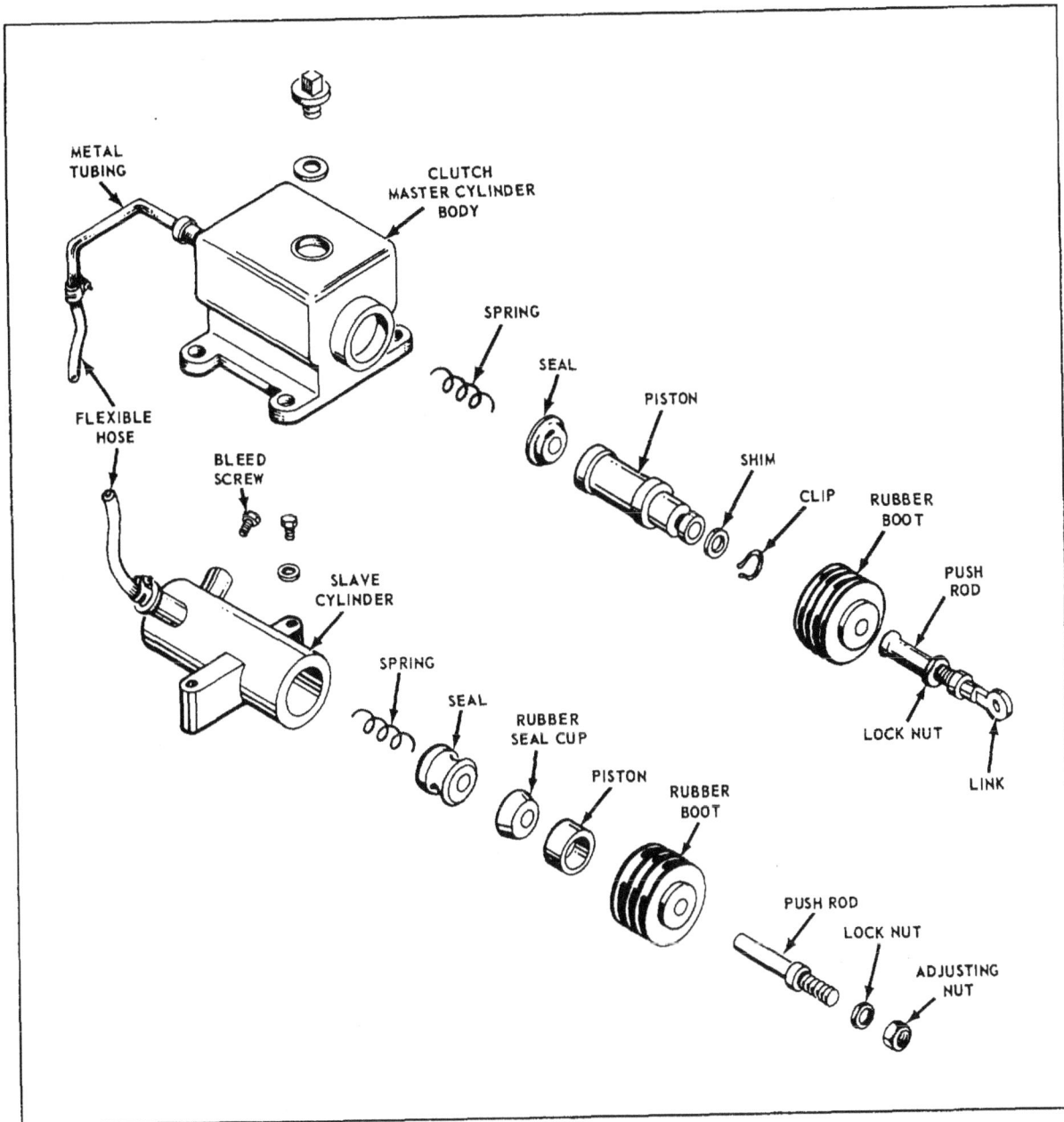

Figure 13-3.-Master cylinder, slave cylinder, and connections for standard hydraulic clutch.

clutch. The plate clutch is a simple clutch with three plates, one of which is clamped between the other two. Figure 13-2 shows exploded and cross-sectional views of a plate clutch.

SINGLE-DISK CLUTCH

The driving members of the single-disk clutch consist of the flywheel and the driving (pressure) plate. The driven member consists of a single disk, splined to the clutch shaft and faced on both sides with friction material. When the clutch is fully engaged, the driven disc is firmly clamped between the flywheel and the driving plate by the pressure of the clutch springs. That results in a direct, nonslipping connection between the driving and driven members of the clutch. In this position, the driven disc rotates the clutch shaft to which it is splined. The clutch shaft is connected to the driving wheels through the transmission, propeller shaft, final drive, differential, and live axles.

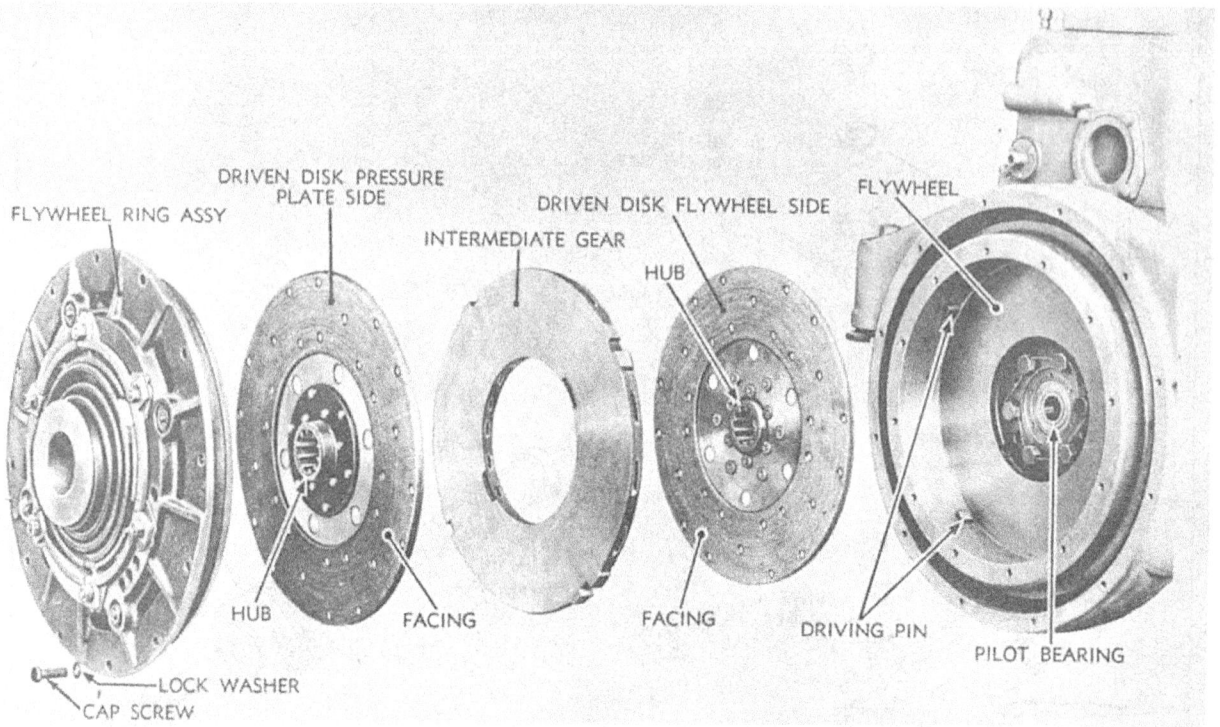

Figure 13-4.-Double-disk clutch-exploded view.

FLYWHEEL RING ASSY

DRIVEN DISK PRESSURE PLATE SIDE

INTERMEDIATE GEAR

DRIVEN DISK FLYWHEEL SIDE

HUB

FLYWHEEL

HUB

FACING

FACING

DRIVING PIN

PILOT BEARING

LOCK WASHER

CAP SCREW

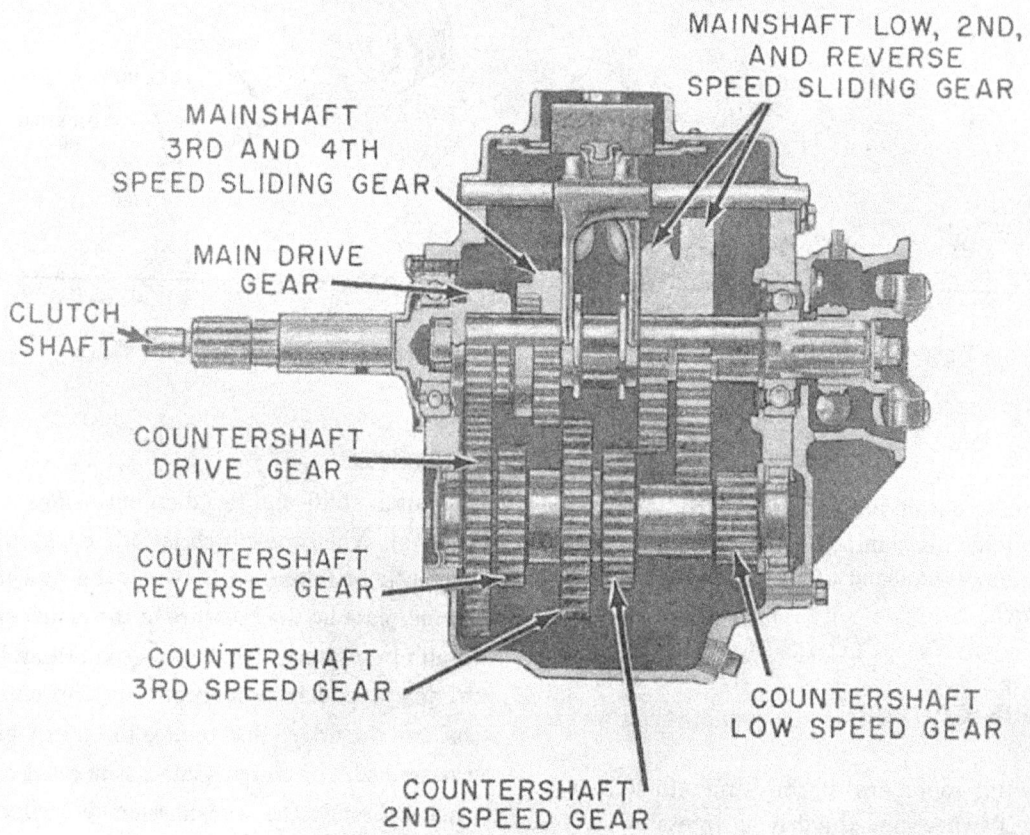

Figure 13-5.-Four-speed truck transmission.

MAINSHAFT LOW, 2ND, AND REVERSE SPEED SLIDING GEAR

MAINSHAFT 3RD AND 4TH SPEED SLIDING GEAR

MAIN DRIVE GEAR

CLUTCH SHAFT

COUNTERSHAFT DRIVE GEAR

COUNTERSHAFT REVERSE GEAR

COUNTERSHAFT 3RD SPEED GEAR

COUNTERSHAFT 2ND SPEED GEAR

COUNTERSHAFT LOW SPEED GEAR

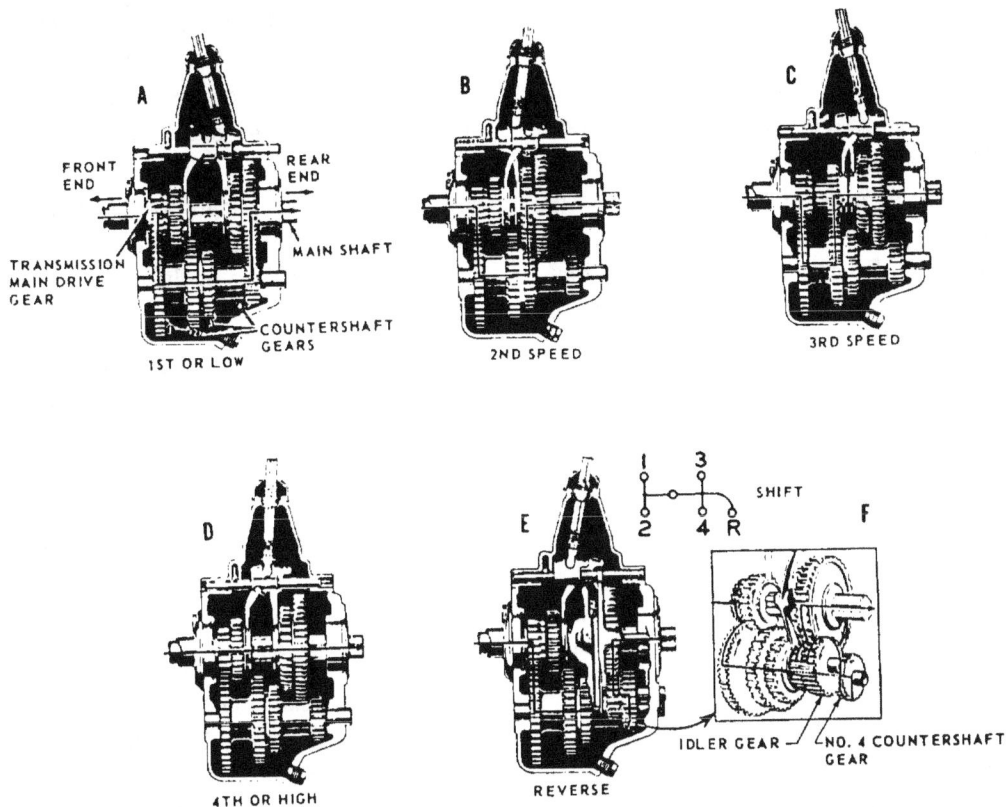

Figure 13-6.-Power flow through a four-speed transmission.

The double-disk clutch (fig. 13-4) is basically the same as the single-plate disk clutch except that another driven disk and intermediate driving plate are added.

MULTIPLE-DISK CLUTCH

A multiple-disk clutch is one having more than three plates or disks. Some have as many as 11 driving plates and 10 driven disks. Because the multiple-disk clutch has a greater frictional area than a plate clutch, it is suitable as a steering clutch on crawler types of tractors. The multiple-disk clutch is sometimes used on heavy trucks. Its operation is very much like that of the plate clutch and has the same release mechanism. The facings, however, are usually attached to the driving plates rather than to the driven disks. That reduces the weight of the driven disks and keeps them from spinning after the clutch is released.

You may run into other types of friction clutches such as the lubricated plate clutch and the cone clutch. These types are seldom used on automatic equipment. However, fluid drives are largely replacing the friction clutches in automobiles, light trucks, and some tractors.

For information on fluid drives (automatic transmissions), refer to *Construction Mechanic 3 & 2,* NAVPERS 10644G-1, chapter 11.

TRANSMISSION

The transmission is part of the power train. It consists of a metal case filled with gears (fig. 13-5). It is usually located in the rear of the engine between the clutch housing and the propeller shaft, as shown in figure 13-1. The transmission transfers engine power from the clutch shaft to the propeller shaft. It allows the driver or operator to control the power and speed of the vehicle. The transmission shown in figure 13-5 and 13-6 is a sliding gear transmission. Many late model trucks have either a constant mesh or synchromesh transmission (explained later). However, both transmissions have the same principles of operation and the same gear ratios.

A review of chapter 6 of this book will help you to understand the transmissions and power transfer mechanisms described in this chapter.

FOUR-SPEED TRUCK TRANSMISSION

The gear shift lever positions shown in the small inset in figure 13-6 are typical of most four-speed truck transmissions. The gear shifting lever, shown in A, B, C, D, and E of the figure, moves the position of the two shifting forks that slide on separate shafts secured in the transmission case cover. Follow the separate diagrams to learn what takes place in shifting from one speed to another. For example, as you move the top of the gear shift toward the forward left position, the lower arm of the lever moves in the opposite direction to shift the gears. The fulcrum of this lever is in the transmission cover.

Shifting transmission gears requires the use of the clutch to disengage the engine. Improper use of the clutch will cause the gears to clash and may damage them by breaking the gear teeth. A broken tooth or piece of metal can wedge itself between two moving gears and ruin the entire transmission assembly.

When you shift from neutral to first, or low, speed (fig. 13-6, A), the smallest countershaft gear engages with the large sliding gear. Low gear moves the truck at its lowest speed and maximum power. The arrows show the flow of power from the clutch shaft to the propeller shaft.

The second-speed position is obtained by moving the gear shift lever straight back from the low-speed position. You will, of course, use the clutch when shifting. In figure 13-6, B, you will see that the next to the smallest countershaft gear is in mesh with the second largest sliding gear. The largest sliding gear (shift gear) has been disengaged, The flow of power has been changed as shown by the arrow. The power transmitted to the wheels in second gear (speed) is less, but the truck will move at a greater speed than it will in low gear if the engine speed is kept the same.

In shifting from the second-speed to the third-speed position, you move the gear shift lever through the neutral position. You must do that in all selective gear transmissions. From the neutral position the driver can select the speed position required to get the power needed. In figure 13-6, C, notice that the gear shift lever is in contact with the other shifting fork and that the forward sliding gear meshes with the second countershaft gear. The power flow through the transmission has again been changed, as indicated by the arrow, and the truck will move at an intermediate speed between second and high.

You shift into the fourth, or high-speed, position by moving the top of the shift lever back and to the right from the neutral position. In the high-speed position, the forward shift or sliding gear is engaged with the constant speed gear as shown in figure 13-6, D. The clutch shaft and the transmission shaft are now locked together, and the power flow is in a straight line. In high, the truck propeller shaft revolves at the same speed as the engine crankshaft, or at a 1 to 1 ratio.

You shift to reverse by moving the top of the gear shift lever to the far right and then to the rear. Most trucks have a trigger arrangement at the gear shift ball to unlock the lever so that it can be moved from neutral to the far right. The lock prevents unintentional shifts into reverse. Never try to shift into reverse until the forward motion of the vehicle has been completely stopped.

In figure 13-6, F, you can see how the idler gear fits into the transmission gear train. In figure 13-6, E, you can see what happens when you shift into reverse. An additional shifting fork is contacted by the shift lever in the far right position. When you shift to reverse, this fork moves the idling gear into mesh with the small countershaft gear and the large sliding gear at the same time. The small arrows in the inset show how the engine power flows through the transmission to move the propeller shaft and the wheels in a reverse direction.

The different combination of gears in the transmission case makes it possible to change the vehicle speed while the engine speed remains the same. It is all a matter of gear ratios. That is, having large gears drive small gears, and having small gears drive large gears. If a gear with 100 teeth drives a gear with 25 teeth, the small gear will travel four times as fast as the large one. You have stepped up the speed. Now, let the small gear drive the large gear, and the large gear will make one revolution for every four of the small gear. You have reduced speed, and the ratio of gear reduction is 4 to 1.

In the truck transmission just described, the gear reduction in low gear is 7 to 1 from the engine to the propeller shaft. In high gear the ratio is 1 to 1, and the propeller shaft turns at the same speed as the engine. This principle holds true for most transmissions. The second- and third-speed positions provide intermediate gear reductions between low and high. The gear ratio in second speed is 3.48 to 1, and in third is 1.71 to 1. The gear reduction or gear ratio in reverse is about the same as it is in low gear, and the propeller shaft makes one revolution for every seven revolutions of the engine.

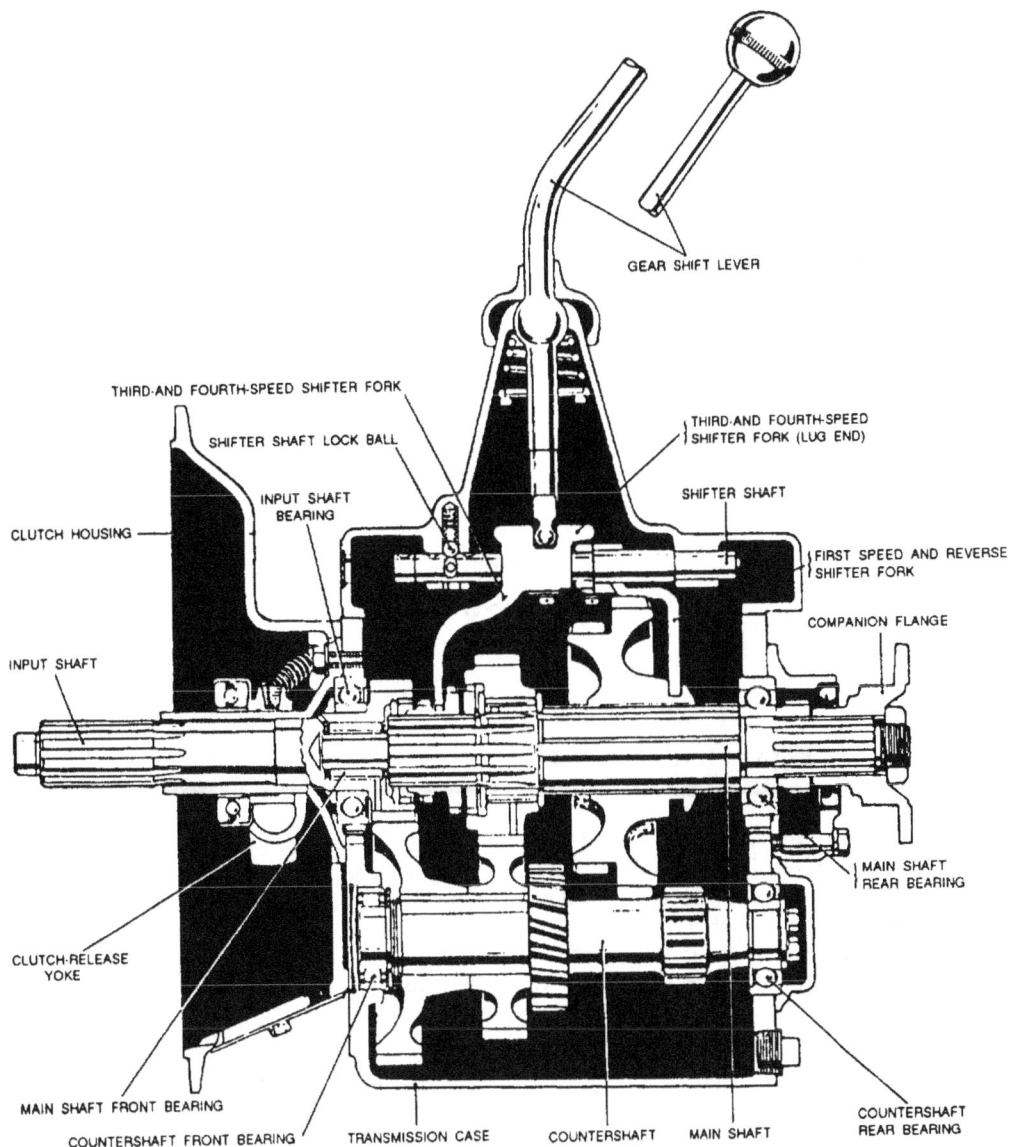

Figure 13-7.-Constant-mesh transmission assembly—sectional view.

All transmissions do not have four speeds forward, and all do not have the same gear reductions at the various speeds. Passenger cars, for example, usually have only three forward speeds and one reverse speed. Their gear ratios are about 3 to 1 in both low and reverse gear combinations. You must remember, the gear reduction in the transmission is only between the engine and the propeller shaft. Another reduction gear ratio is provided in the rear axle assembly. If you have a common rear axle ratio of about 4 to 1, the gear reduction from the engine of a passenger car to the rear wheels in low gear would be approximately 12 to 1. In high gear the ratio would be 4 to 1 since the transmission would have no reduction of speed.

CONSTANT MESH TRANSMISSION

To eliminate the noise developed in the old spur-tooth type of gears used in the sliding gear transmission, the automotive manufacturers developed the constant-mesh transmission that contains helical gears.

In this type of transmission, certain countershaft gears are constantly in mesh with the main shaft gears. The main shaft meshing gears are arranged so that they cannot move endwise. They are supported by roller bearings that allow them to rotate independently of the main shaft (figs. 13-7 and 13-8).

In operation, when you move the shift lever to third, the third and fourth shifter fork moves the clutch gear

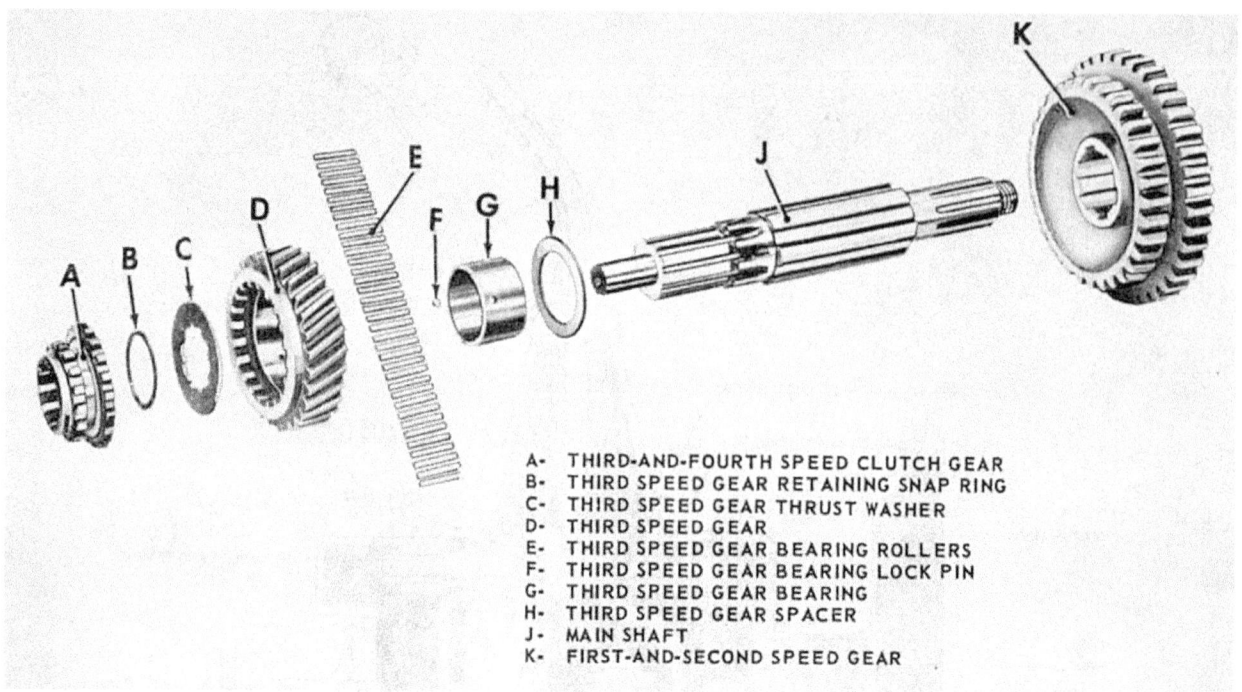

A- THIRD-AND-FOURTH SPEED CLUTCH GEAR
B- THIRD SPEED GEAR RETAINING SNAP RING
C- THIRD SPEED GEAR THRUST WASHER
D- THIRD SPEED GEAR
E- THIRD SPEED GEAR BEARING ROLLERS
F- THIRD SPEED GEAR BEARING LOCK PIN
G- THIRD SPEED GEAR BEARING
H- THIRD SPEED GEAR SPACER
J- MAIN SHAFT
K- FIRST-AND-SECOND SPEED GEAR

Figure 13-8.—Dissembled main shaft assembly.

(fig. 13-8, A) toward the third-speed gear (fig. 13-8, D). This action engages the external teeth of the clutch gear with the internal teeth of the third-speed gear. Since the third-speed gear is rotating with the rotating counter-shaft gear, the clutch gear also must rotate. The clutch gear is splined to the main shaft, and therefore, the main shaft rotates with the clutch gear. This principle is carried out when the shift lever moves from one speed to the next.

Constant-mesh gears are seldom used for all speeds. Common practice is to use such gears for the higher gears, with sliding gears for first and reverse speeds, or for reverse only. When the shift is made to first or reverse, the first and reverse sliding gear is moved to the left on the main shaft. The inner teeth of the sliding gear mesh with the main shaft first gear.

SYNCHROMESH TRANSMISSION

The synchromesh transmission is a type of constant-mesh transmission. It synchronizes the speeds of mating parts before they engage to allow the selection of gears without their clashing. It employs a combination metal-to-metal friction cone clutch and a dog or gear positive clutch. These clutches allow the main drive gear and second-speed main shaft gear to engage with the transmission main shaft. The friction cone clutch engages first, bringing the driving and driven members to the same speed, after which the dog clutch engages easily without clashing. This process is accomplished in one continuous operation when the driver declutches and

moves the control lever in the usual manner. The construction of synchromesh transmissions varies somewhat with different manufacturers, but the principle is the same in all.

The construction of a popular synchromesh clutch is shown in figure 13-9. The driving member consists of a sliding gear splined to the transmission main shaft with bronze internal cones on each side. It is surrounded by a sliding sleeve having internal teeth that are meshed with the external teeth of the sliding gear. The sliding sleeve has grooves around the outside to receive the shift fork. Six spring-loaded balls in radially drilled holes in the gear fit into an internal groove in the sliding sleeve. That prevents the sliding sleeve from moving endwise relative to the gear until the latter has reached the end of its travel. The driven members are the main drive gear and second-speed main shaft gear. Each has external cones and external teeth machined on its sides to engage the internal cones of the sliding gear and the internal teeth of the sliding sleeve.

The synchromesh clutch operates as follows: when the driver moves the transmission control lever to the third-speed, or direct-drive, position the shift fork moves the sliding gear and sliding sleeve forward as a unit until the internal cone on the sliding gear engages the external cone on the main drive gear. This action brings the two gears to the same speed and stops endwise travel of the sliding gear. The sliding sleeve slides over the balls and silently engages the external teeth on the main drive gear. This action locks the main drive gear and transmission main shaft together as shown in

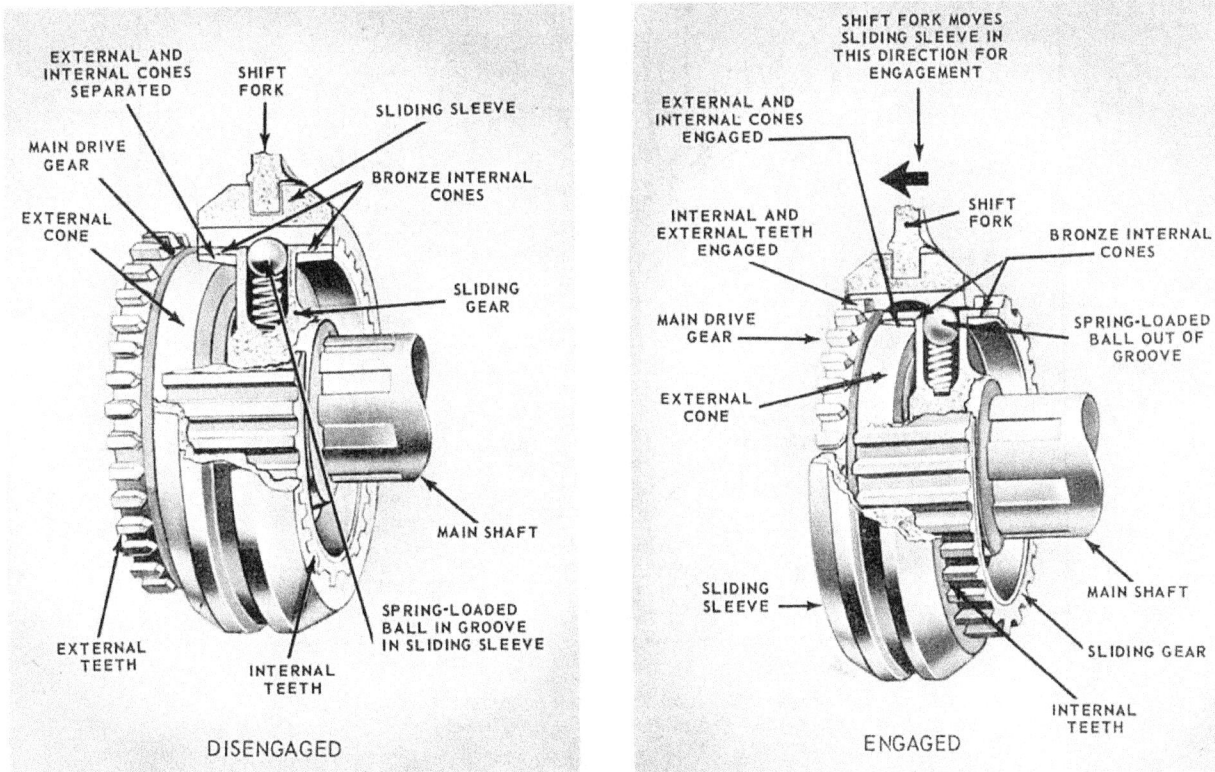

Figure 13-9.-Synchromesh clutch-disengaged and engaged.

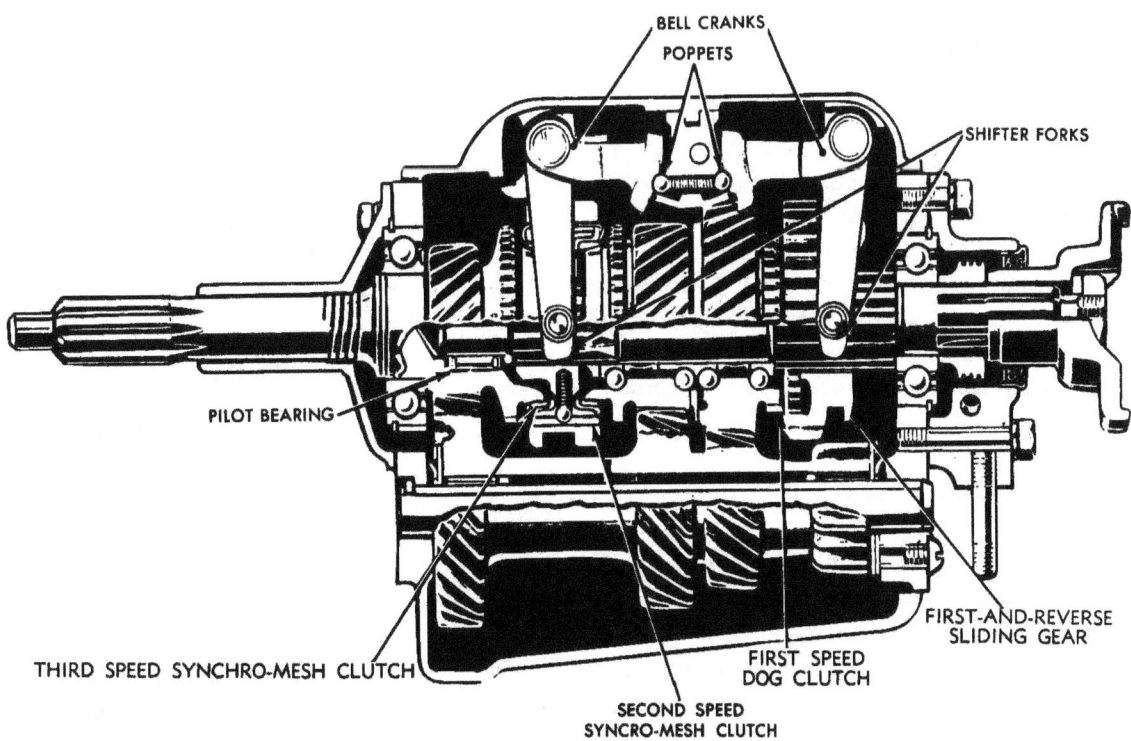

Figure 13-10.-Synchromesh transmission arranged for steering column control.

figure 13-9. When the transmission control lever is shifted to the second-speed position, the sliding gear and sleeve move rearward. The same action takes place, locking the transmission main shaft to the second-speed

main shaft gear. The snchromesh clutch is not applied to first speed or to reverse. First speed is engaged by an ordinary dog clutch when constant mesh is employed by a sliding gear. Figure 13-10 shows a cross section of a

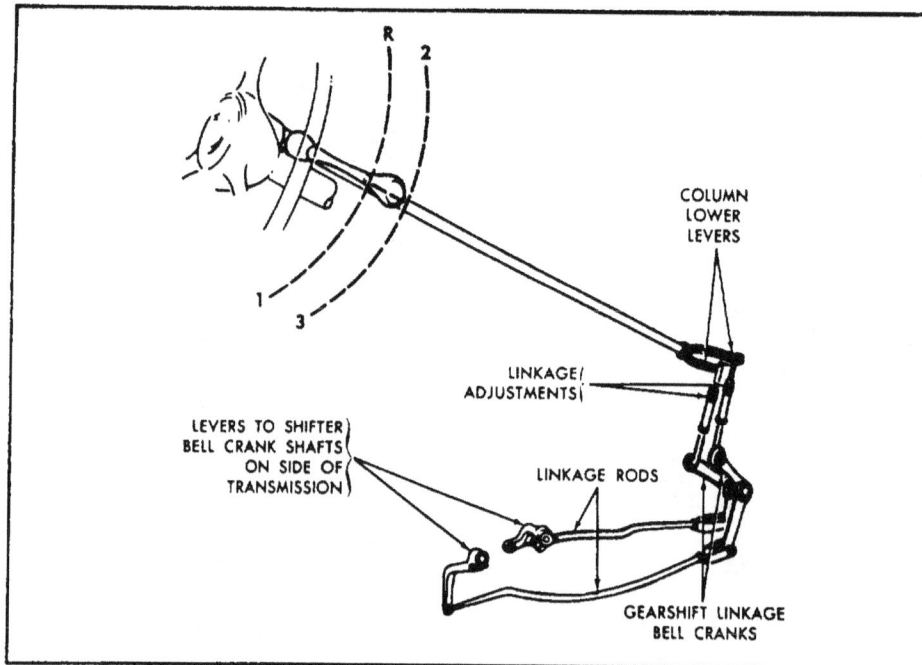

Figure 13-11.-Steering colunn transmission control lever and linkage.

synchromesh transmission that uses constant-mesh helical gears for the three forward speeds and a sliding spur gear for reverse.

Some transmissions are controlled by a steering column control lever (fig. 13-11). The positions for the various speeds are the same as those for the vertical control lever except that the lever is horizontal. The shifter fork is pivoted on bell cranks that are turned by a steering column control lever through the linkage shown. The poppets shown in figure 13-10 engage notches at the inner end of each bell crank. Other types of synchromesh transmissions controlled by steering column levers have shifter shafts and forks moved by a linkage similar to those used with a vertical control lever.

AUXILIARY TRANSMISSION

The auxiliary transmission allows a rather small truck engine to move heavy loads by increasing the engine-to-axle gear ratios. The auxiliary transmission provides a link in the power trains of construction vehicles. This link diverts engine power to drive four and six wheels and to operate accessory equipment through transfer cases and power takeoff units. (See fig. 13-12).

Trucks require a greater engine-to-axle gear ratio than passenger cars, particularly when manufacturers put the same engine in both types of equipment. In a truck, the auxiliary transmission doubles the mechanical advantage. It connects to the rear of the main transmission by a short propeller shaft and universal joint. Its weight is supported on a frame crossmember as shown in figure 13-12. The illustration also shows how the shifting lever would extend into the driver's compartment near the lever operating the main transmission.

In appearance and in operation, auxiliary transmissions are similar to main transmissions, except that some may have two and some three speeds (low, direct, and overdrive).

TRANSFER CASES

Transfer cases are put in the power trains of vehicles driven by all wheels. Their purpose is to provide the necessary offsets for additional propeller shaft connections to drive the wheels.

Transfer cases in heavier vehicles have two speed positions and a declutching device for disconnecting the front driving wheels. Two speed transfer cases, such as the one shown in figure 13-13, serve also as auxiliary transmissions.

Some transfer cases are complicated. When they have speed-changing gears, declutching devices, and attachments for three or more propeller shafts, they are even larger than the main transmission. A cross section

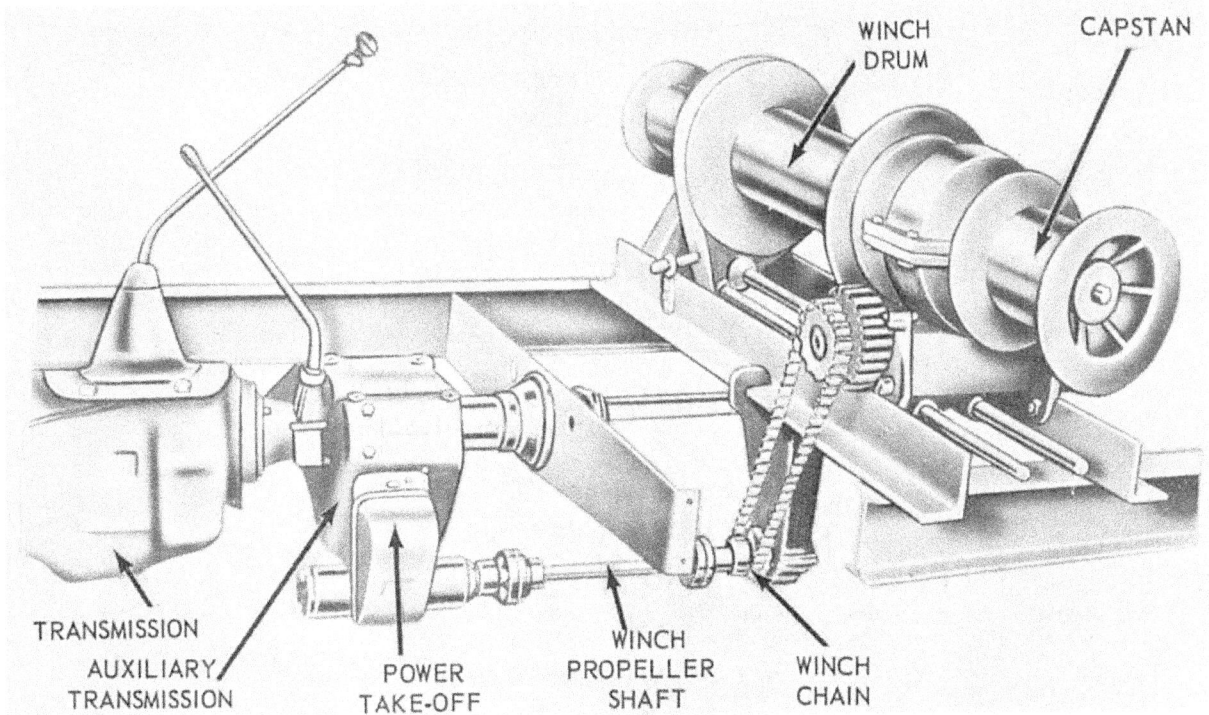

Figure 13-12.—Auxiliary transmission power takeoff driving winch.

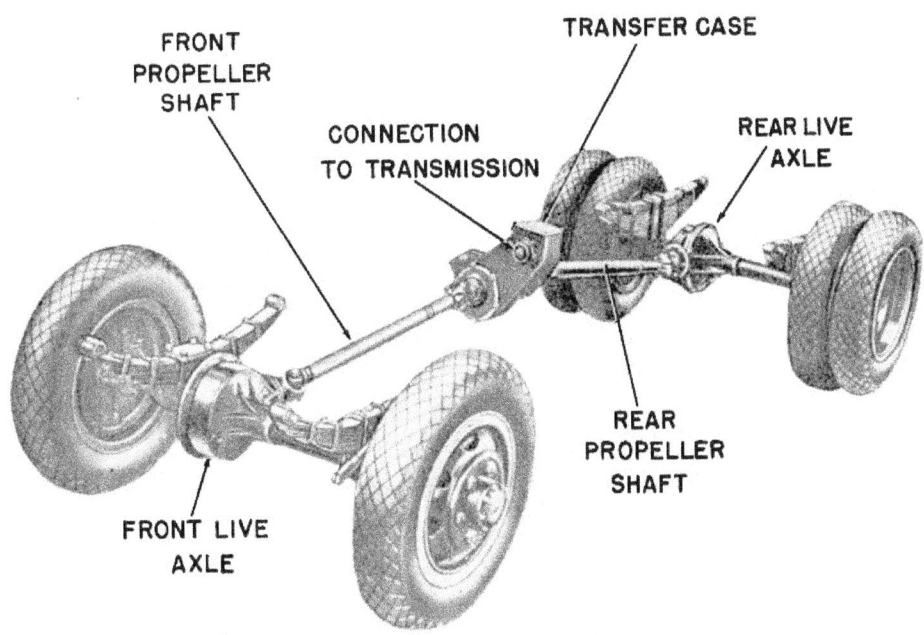

Figure 13-13.—Transfer case installed in a four-wheel drive truck.

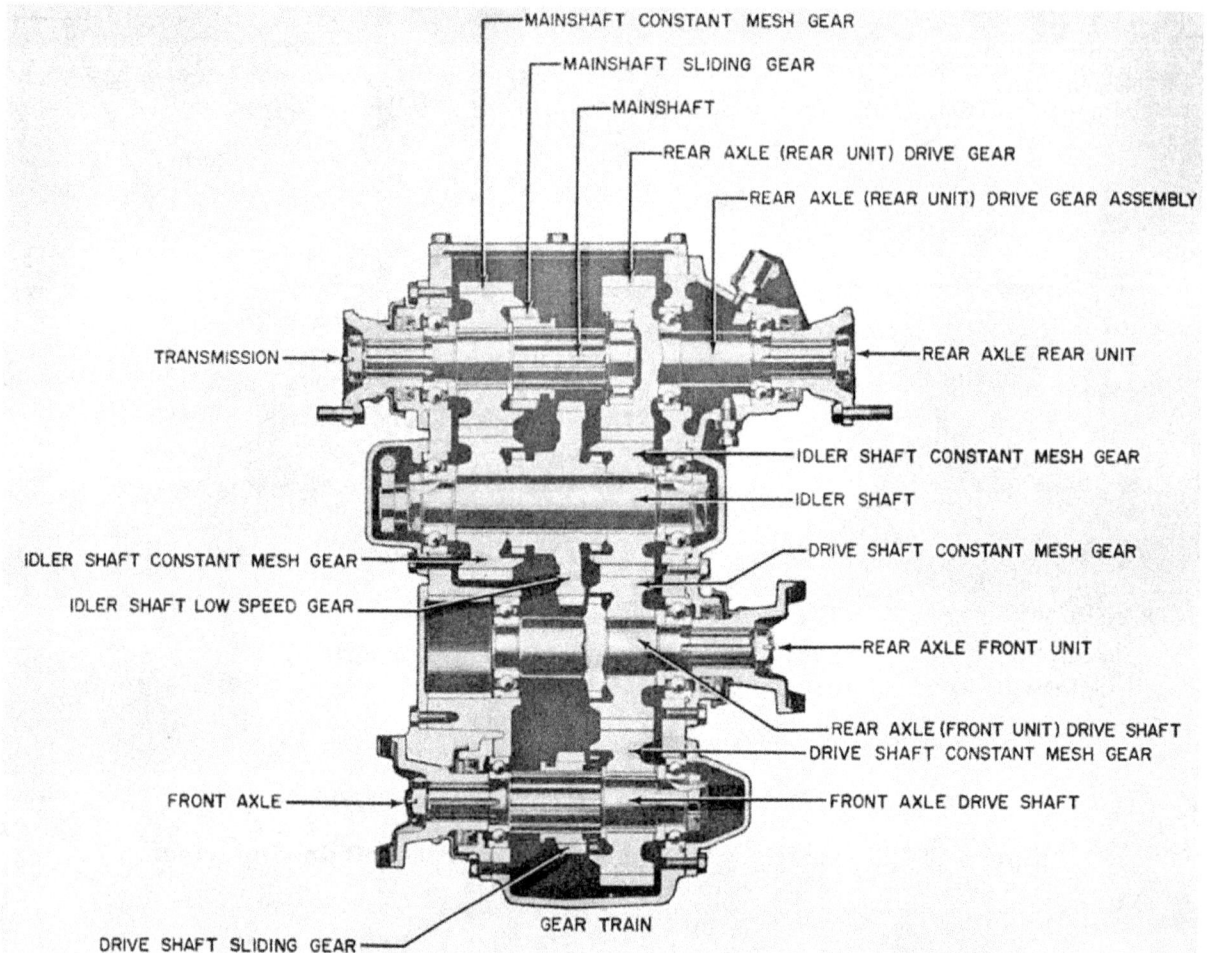

Figure 13-14.-Cross section of a two-speed transfer case.

of a common type of two-speed transfer case is shown in figure 13-14. Compare it with the actual installation in figure 13-13.

This same type of transfer case is used for a six-wheel drive vehicle. The additional propeller shaft connects the drive shaft of the transfer case to the rearmost axle assembly. It is connected to the transfer case through the transmission brake drum.

Some transfer cases contain an overrunning sprag unit (or units) on the front output shaft. (A sprag unit is a form of overrunning clutch; power can be transmitted through it in one direction but not in the other.)

On these units the transfer is designed to drive the front axle slightly slower than the rear axle. During normal operation, when both front and rear wheels turn at the same speed, only the rear wheels should lose traction and begin to slip. They tend to turn faster than the front wheels. As slipping occurs, the sprag unit automatically

engages so that the front wheels also drive the vehicle. The sprag unit simply provides an automatic means of engaging the front wheels in drive whenever additional tractive effort is required. There are two types of sprag-unit-equipped transfers, a single-sprag unit transfer and a double-sprag unit transfer. Essentially, both types work in the same manner.

POWER TAKEOFFS

Power takeoffs are attachments in the power train for power to drive auxiliary accessories. They are attached to the transmission, auxiliary transmission, or transfer case. A common type of power takeoff is the single-gear, single-speed type shown in figure 13-15. The unit bolts to an opening provided in the side of the transmission case as shown in figure 13-12. The sliding gear of the power takeoff will then mesh with the transmission countershaft gear. The operator can move a shifter shaft control lever to slide the gear in and out

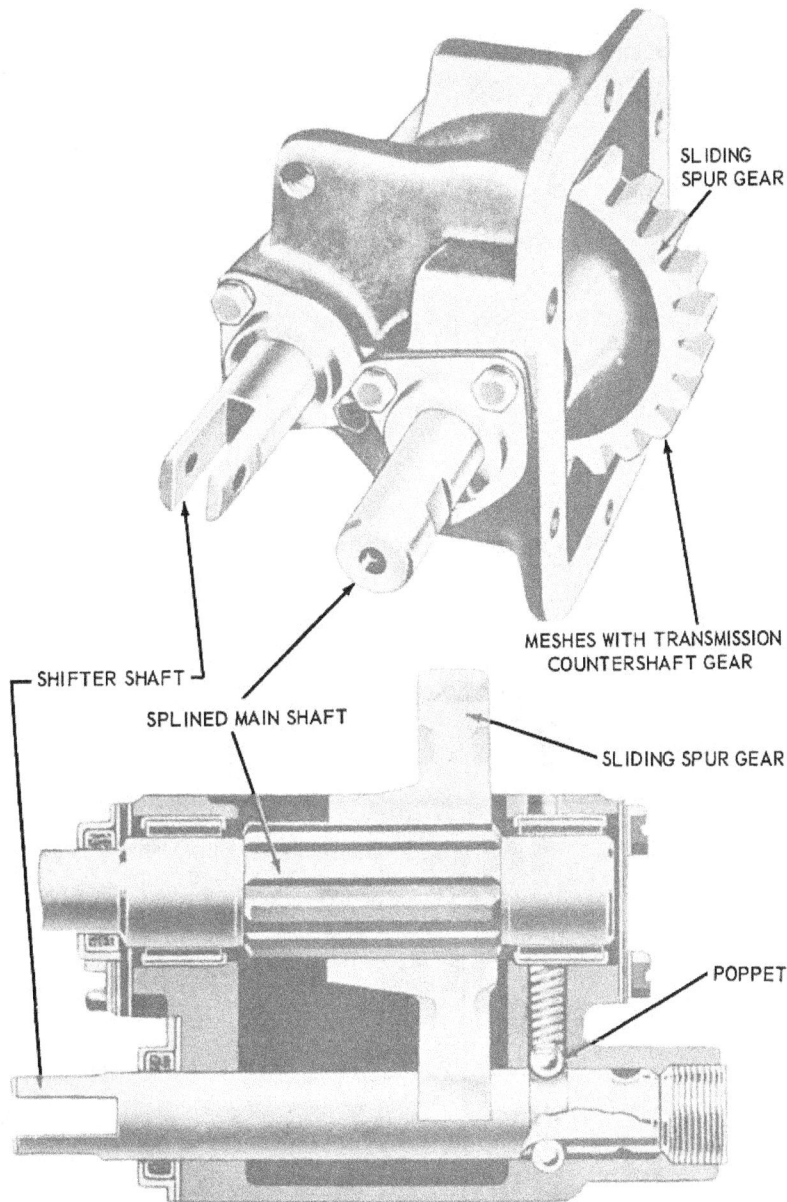

Figure 13-15.-Single-speed, single-gear, power takeoff.

of mesh with the countershaft gear. The spring-loaded ball holds the shifter shaft in position.

On some vehicles you will find power take-off units with gear arrangements that will give two speeds forward and one in reverse. Several forward speeds and a reverse gear arrangement are usually provided in power take-off units that operate winches and hoists. Their operation is about the same as that in the single-speed units.

PROPELLER SHAFT ASSEMBLIES

The propeller shaft assembly consists of a propeller shaft, a slip joint, and one or more universal joints. This assembly provides a flexible connection through which power is transmitted from the transmission to the live axle.

The propeller shaft may be solid or tubular. A solid shaft is stronger than a hollow or tubular shaft of the same diameter, but a hollow shaft is stronger than a solid shaft of the same weight. Solid shafts are used inside a shaft housing that encloses the entire propeller shaft assembly. These are called torque tube drives.

A slip joint is put at one end of the propeller shaft to take care of end play. The driving axle, attached to the springs, is free to move up and down, while the transmission is attached to the frame and cannot move. Any

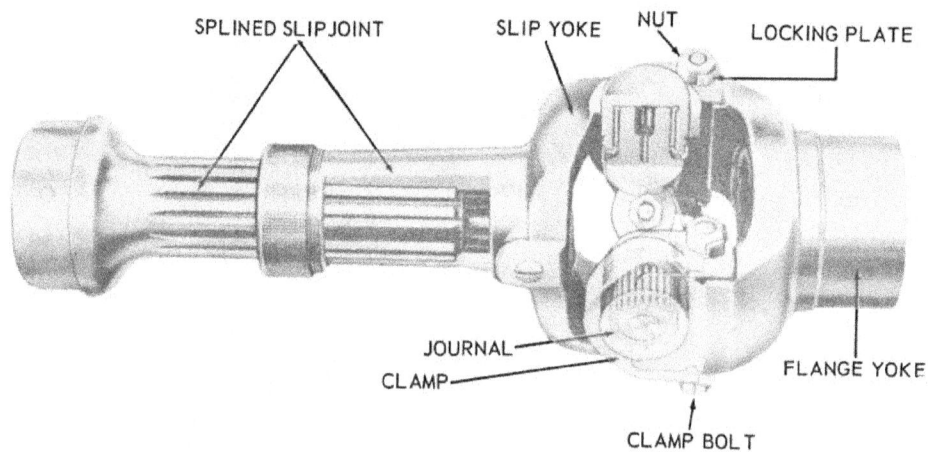

Figure 13-16.—Slip joint and common type of universal Joint.

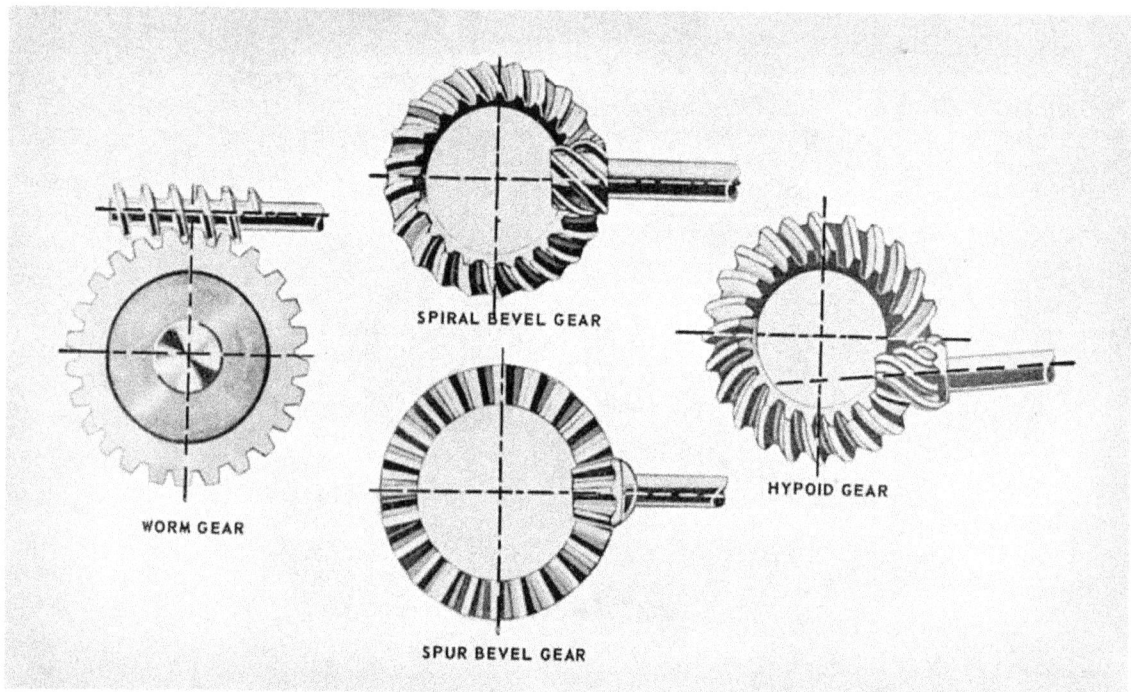

Figure 13-17.—Gears used in final drives.

upward or downward movement of the axle, as the springs flex, shortens or lengthens the distance between the axle assembly and the transmission. This changing distance is compensated for by a slip joint placed at one end of the propeller shaft.

The usual type of slip joint consists of a splined stub shaft, welded to the propeller shaft, that fits into a splined sleeve in the universal joint. A slip joint and universal joint are shown in figure 13-16.

Universal joints are double-hinged with the pins of the hinges set at right angles. They are made in

many different designs, but they all work on the same principle. (See chapter 11.)

FINAL DRIVES

A final drive is that part of the power train that transmits the power delivered through the propeller shaft to the drive wheels or sprockets. Because it is encased in the rear axle housing, the final drive is usually referred to as a part of the rear axle assembly. It consists of two gears called the ring gear and pinion. These may be spur, spiral, hypoid beveled, or worm gears, as illustrated in figure 13-17.

The function of the final drive is to change by 90 degrees the direction of the power transmitted through the propeller shaft to the driving axles. It also provides a fixed reduction between the speed of the propeller shaft and the axle shafts and wheels. In passenger cars this reduction varies from about 3 to 1 to 5 to 1. In trucks, it can vary from 5 to 1 to as much as 11 to 1.

The gear ratio of a final drive having bevel gears is found by dividing the number of teeth on the drive gear by the number of teeth on the pinion. In a worm gear final drive, you find the gear ratio by dividing the number of teeth on the gear by the number of threads on the worm.

Most final drives are of the gear type. Hypoid gears (fig. 13-17) are used in passenger cars and light trucks to give more body clearance. They permit the bevel drive pinion to be put below the center of the bevel drive gear, thereby lowering the propeller shaft. Worm gears allow a large speed reduction and are used extensively in larger trucks. Spiral bevel gears are similar to hypoid gears. They are used in both passenger cars and trucks to replace spur gears that are considered too noisy.

DIFFERENTIALS

Chapter 11 described the construction and principle of operation of the gear differential. We will briefly review some of the high points of that chapter here and describe some of the more common types of gear differentials applied in automobiles and trucks.

The purpose of the differential is easy to understand when you compare a vehicle to a company of sailors marching in mass formation. When the company makes a turn, the sailors in the inside file must take short steps, almost marking time, while those in the outside file must take long steps and walk a greater distance to make the turn. When a motor vehicle turns a comer, the wheels outside of the turn must rotate faster and travel a greater distance than the wheels on the inside. That causes no difficulty for front wheels of the usual passenger car because each wheel rotates independently on opposite ends of a dead axle. However, to drive the rear wheel at different speeds, the differential is needed. It connects the individual axle shaft for each wheel to the bevel drive gear. Therefore, each shaft can turn at a different speed and still be driven as a single unit. Refer to the illustration in figure 13-18 as you study the following discussion on differential operation.

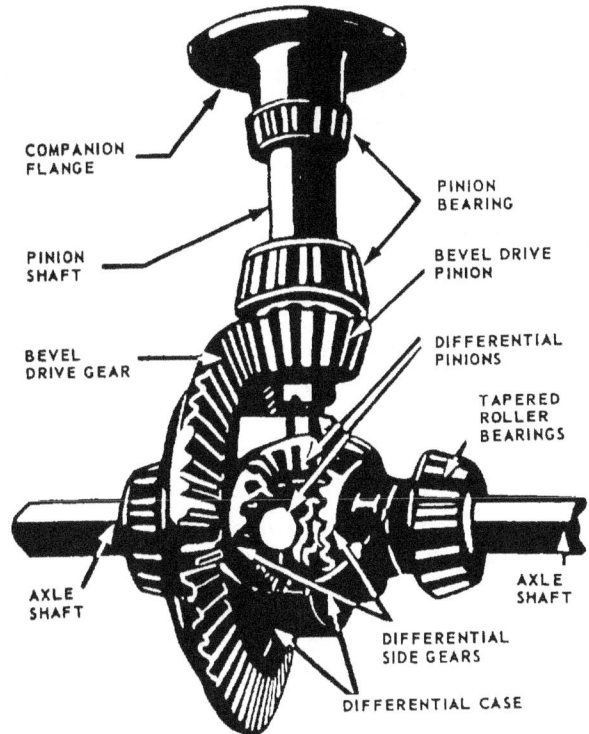

Figure 13-18.-Differential with part of case cut away.

The differential described in chapter 11 had two inputs and a single output. The differential used in the automobile has a single input and two outputs. Its input is introduced from the propeller shaft and its outputs goes to the rear axles and wheels.

The bevel drive pinion, connected to the pinion shaft, drives the bevel drive gear and the differential case to which it is attached. Therefore, the entire, differential case always rotates with the bevel drive gear whenever the pinion shaft is transmitting rotary motion. Within the case, the differential pinions (refereed to as spider gears in chapter 11) are free to rotate on individual shafts called trunnions. These trunnions are attached to the walls of the differential case. Whenever the case is turning, the differential pinions must revolve-one about the other-in the same plane as the bevel drive gear.

The differential pinions mesh with the side gears, as did the spider and side gears in the differential described in chapter 11. The axle shafts are splined to the differential side gears and keyed to the wheels. Power is transmitted to the axle shafts through the differential pinions and the side gears. When resistance is equal on each rear wheel, the differential pinions, side gears, and axle shafts all rotate as one unit with the bevel drive gear. In this case, there is no relative motion between the pinions and the side gears in the differential case. That is, the pinions do not turn on the trunnions, and their teeth will not move over the teeth of the side gears.

When the vehicle turns a comer, one wheel must turn faster than the other. The side gear driving the outside wheel will run faster than the side gear connected to the axle shaft of the inside wheel. To compensate for this difference in speed and to remain in mesh with the two side gears, the differential pinions must then turn on the trunnions. The average speed of the two side gears, axle shafts, or wheels is always equal to the speed of the bevel drive gear.

Some trucks are equipped with a differential lock to prevent one wheel from spinning. This lock is a simple dog clutch, controlled manually or automatically, that locks one axle shaft to the differential case and bevel drive gear. This device forms a rigid connection between the two axle shafts and makes both wheels rotate at the same speed. Drivers seldom use it, however, because they often forget to disengage the lock after using it.

Several automotive devices are available that do almost the same thing as the differential lock. One that is used extensively today is the high-traction differential. It consists of a set of differential pinions and side gears that have fewer teeth and a different tooth form from the conventional gears. Figure 13-19 shows a comparison between these and standard gears.

The high-traction differential pinions and side gears depend on a variable radius from the center of the differential pinion to the point where it comes in contact with the side gear teeth, which is, in effect, a variable lever arm. While there is relative motion between the pinions and side gears, the torque is unevenly divided between the two driving shafts and wheels; whereas, with the usual differential, the torque is evenly divided always. With the high-traction differential, the torque becomes greater on one wheel and lesson the other as the pinions move around, until both wheels start to rotate at the same speed. When that occurs, the relative motion between the pinion and side gears stops and the torque on each wheel is again equal. This device helps to start the vehicle or keep it rolling when one wheel encounters a slippery spot and loses traction while the other wheel is on a firm spot and has traction. It will not work however, when one wheel loses traction completely. In this respect, it is inferior to the differential lock.

With the no-spin differential (fig. 13-20), one wheel cannot spin because of loss of tractive effort and thereby deprive the other wheel of driving effort. For example, one wheel is on ice and the other wheel is on dry pavement. The wheel on ice is assumed to have no traction. However, the wheel on dry pavement will pull to the limit of its tractional resistance at the pavement. The wheel on ice cannot spin because wheel speed is

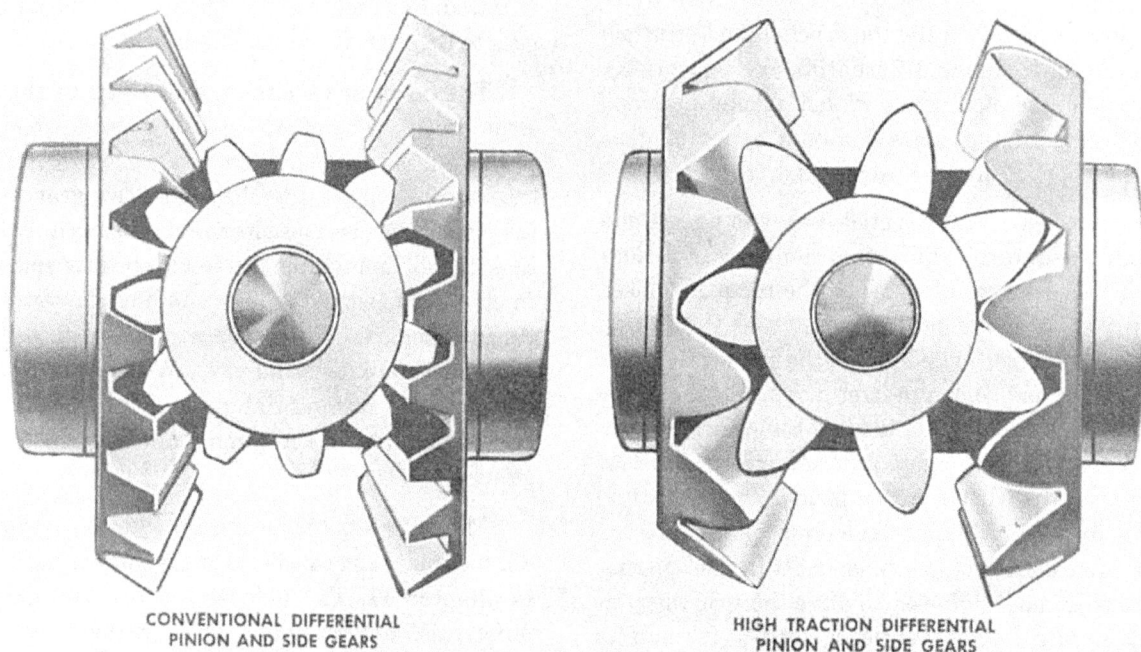

CONVENTIONAL DIFFERENTIAL
PINION AND SIDE GEARS

HIGH TRACTION DIFFERENTIAL
PINION AND SIDE GEARS

Figure 13-19.-Comparison of high-traction differential gears and standard differential gears.

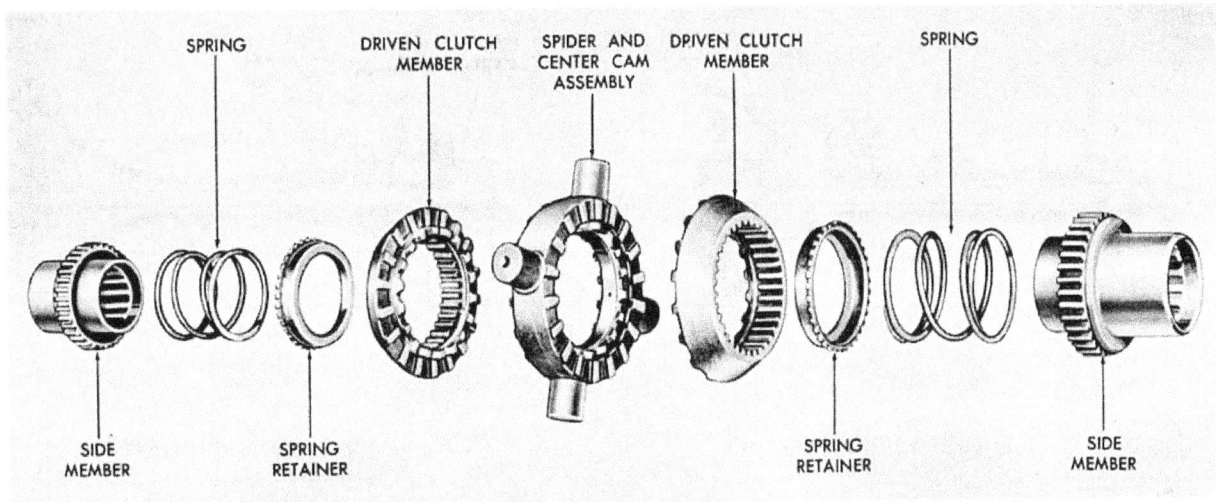

Figure 13-20.—No spin differential—exploded view.

governed by the speed of the wheel applying tractive effort.

The no-spin differential does not contain pinion gears and side gears as does the conventional differential. Instead, it consists basically of a spider attached to the differential drive ring gear through four trunnions. It also has two driven clutch members with side teeth that are indexed by spring pressure with side teeth in the spider. Two side members are splined to the wheel axles and, in turn, are splined into the driven clutch members.

AXLES

A live axle is one that supports part of the weight of a vehicle and drives the wheels connected to it. A dead axle is one that carries part of the weight of a vehicle but does not drive the wheels. The wheels rotate on the ends of the dead axle.

Usually, the front axle of a passenger car is a dead axle and the rear axle is a live axle. In four-wheel drive vehicles, both front and rear axles are live axles; in six-wheel drive vehicles, all three axles are live axles. The third axle, part of a bogie drive, is joined to the rearmost axle by a trunnion axle. The trunnion axle attaches rigidly to the frame. Its purpose is to help distribute the load on the rear of the vehicle to the two live axles that it connects.

Four types of live axles are used in automotive and construction equipment. They are: plain, semifloating, three-quarter floating, and full floating.

The plain live, or nonfloating, rear axle, is seldom used in equipment today. The axle shafts

in this assembly are called nonfloating because they are supported directly in bearings located in the center and ends of the axle housing. In addition to turning the wheels, these shafts carry the entire load of the vehicle on their outer ends. Plain axles also support the weight of the differential case.

The semifloating axle (fig. 13-21) used on most passenger cars and light trucks has its differential case independently supported. The differential carrier relieves the axle shafts from the weight of the differential assembly and the stresses caused by its operation. For this reason the inner ends of the axle shafts are said to be floating. The wheels are keyed to outer ends of axle shafts and the outer bearings are between the shafts and the housing. The axle shafts therefore must take the stresses caused by turning, skidding, or wobbling of the wheels. The axle shaft is a semifloating live axle that can be removed after the wheel has been pulled off.

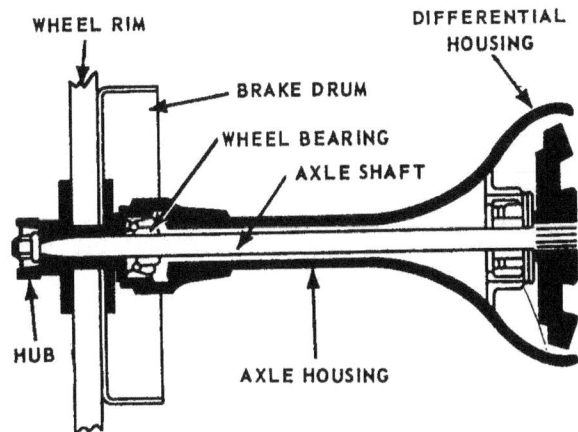

Figure 13-21.—Semifloating rear axle.

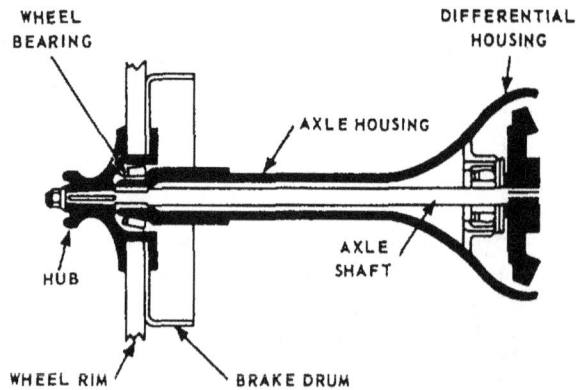

Figure 13-22.-Three-quarter floating rear axle.

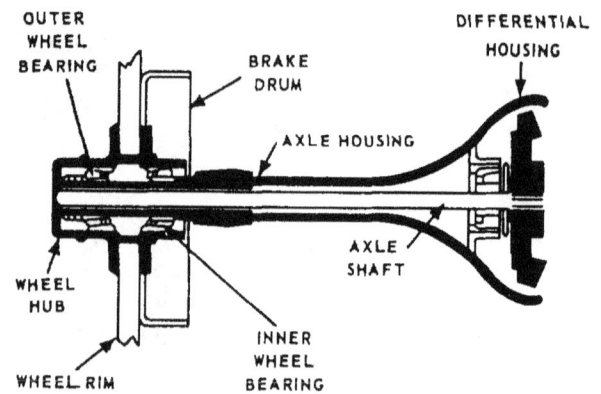

Figure 13-23.-Full floating rear axle.

The axle shafts in a three-quarter floating axle (fig. 13-22) may be removed with the wheels, keyed to the tapered outer ends of the shafts. The inner ends of the shaft are carried as in a semifloating axle. The axle housing, instead of the shafts, carries the weight of the vehicle because the wheels are supported by bearings on the outer ends of the housing. However, axle shafts must take the stresses caused by the turning, skidding, and wobbling of the wheels. Three-quarter floating axles are used in some trucks, but in very few passenger cars. Most heavy trucks have a full floating axle (fig. 13-23). These axle shafts may be removed and replaced without removing the wheels or disturbing the differential. Each wheel is carried on the end of the axle tube on two ball bearings or roller bearings, and the axle shafts are not rigidly connected to the wheels. The wheels are driven through a clutch arrangement or flange on the ends of the axle shaft that is bolted to the outside of the wheel hub. The bolted connection between the axle and wheel does not make this assembly a true full floating axle, but nevertheless, it is called a floating axle. A true full floating axle transmits only turning effort, or torque.

SUMMARY

Chapter 13 explained how power developed by the engine is transmitted to perform the work required of it. It discussed the following mechanisms of the power train:

The clutch is incorporated in the powertrain to provide a means of disconnecting the power of the engine from the driving wheels and accessory equipment.

The transmission transfers engine power from the clutch shaft to the propeller shaft and allows the operator to control the power and speed of the vehicle by selecting various gear ratios.

Transfer cases provide the necessary offsets for additional propeller shaft connections to drive the wheels.

Propeller shaft assemblies provide a flexible connection through which power is transmitted from the transmission to the axle.

Axles are used to support part of the weight of a vehicle; they also drive the wheels connected to them.

APPENDIX I

REFERENCES USED TO DEVELOP THE TRAMAN

Bernstein, Leonard, Martin Schacter, Alan Winkler, and Stanley Wolfe, *Concepts and Challenges in Physical Science,* Cebco Standard Publishing, Fairfield, N.J., 1978.

Construction Mechanics 3 & 2, NAVEDTRA 10644-G1, Naval Education and Training Program Management Support Activity, Pensacola, Fla., 1988.

Eby, Denise, and Robert B. Horton, *Physical Science,* Macmillan Publishing Company, New York, 1988.

Gill, Paul W., James H. Smith, Jr., and Eugene J. Ziurys, *Internal Combustion Engines,* 4th ed., The George Banta Company Inc., Menasha, Wis., 1954.

Harris, Norman C., and Edwin M. Hemmerling, *Introductory Applied Physics,* McGraw-Hill Book Company, Inc., New York, 1955.

Heimler, Charles H., and Jack S. Price, *Focus on Physical Science,* Charles E. Merrill Publishing Company, Columbus, Ohio, 1984.

INDEX

Assignment Questions

Information: The text pages that you are to study are provided at the beginning of the assignment questions.

ASSIGNMENT 1

Textbook   "Levers," chapter 1, pages 1-1 through 1-8; Block and Tackle," chapter
Assignment: 2, pages 2-1 through 2--6; "The Wheel and Axle," chapter 3, pages 3-1
                  through 3-6; "The Inclined Plane and the Wedge," chapter 4, pages: 4-1
                  through 4-2.

1-1. A chain hoist lifts a 300-pound
      load through a height of 10 feet
      because it enables you to lift the
      load by exerting less than 300
      pounds of force over a distance of
      10 feet or less.

      1. True
      2. False

1-2. When a chain hoist is used to
      multiply the force being exerted on
      a load, the chain is pulled at a
      faster rate than the load travels.

      1. True
      2. False

1-3. What are the six basic simple
      machines?

      1. The lever, the block and
          tackle, the inclined plane, the
          engine, the wheel and axle, and
          the gear
      2. The lever, the block and
          tackle, the wheel and axle, the
          screw, the gear, and the
          eccentric
      3. The lever, the block and
          tackle, the wheel and axle, the
          inclined plane, the screw, and
          the gear
      4. The lever, the inclined plane,
          the gear, the screw, the
          fulcrum, and the torque

1-4. Which of the following basic
      principles is recognized by
      physicists as governing each simple
      machine?

      1. The wedge or the screw
      2. The wheel and axle or the gear
      3. The lever or the inclined plane
      4. The block and tackle or the
          wheel and axle

1-5. Which of the following simple
      machines works on the same
      principle as the inclined plane?

      1. Screw
      2. Gear
      3. Wheel and axle
      4. Block and tackle

1-6. The fundamentally important points
      in any lever problem are (1) the
      point at which the force is
      applied, (2) the fulcrum, and (3)
      the point at which the

      1. lever will balance
      2. resistance arm equals the
          effort arm
      3. mechanical advantage begins to
          increase
      4. resistance is applied

1

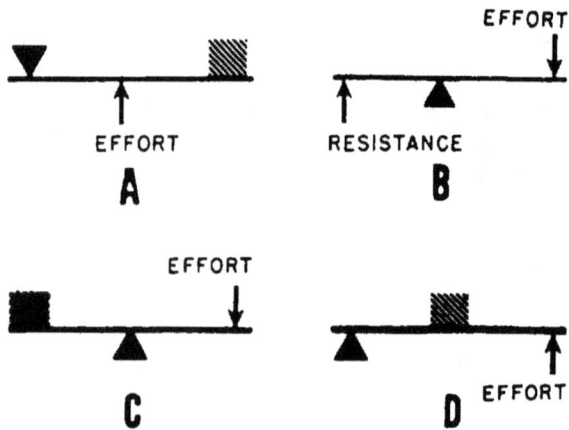

Figure 1A.

QUESTIONS 1-7 THROUGH 1-9 RELATE TO THE DRAWINGS IN FIGURE 1A.

1-7. Which, if any, of the following parts illustrates a first class lever?

1. A
2. B or C
3. D
4. None of the above

1-8. Which part illustrates a second-class lever?

1. D
2. C
3. B
4. A

1-9. What part illustrates a third-class lever?

1. A
2. B
3. C
4. D

1-10. Which of the following classes of levers should you use to lift a large weight by exerting the least effort?

1. First-class
2. Second-class
3. First- or second-class
4. Third-class

1-11. You will find it advantageous to use a third-class lever when the desired result is

1. a transformation of energy
2. an increase in speed
3. a decrease in applied effort
4. a decrease in speed and an increase in applied effort

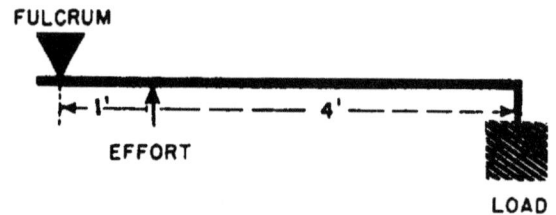

Figure 1B

IN ANSWERING QUESTIONS 1-12 THROUGH 1-14, SELECT THE CORRECT ARM MEASUREMENTS FROM FIGURES 1B AND 1C.

1-12. Effort arm in figure 1B

1. 1 ft
2. 3 ft
3. 4 ft
4. 5 ft

1-13. Resistance arm in figure 1B

1. 1 ft
2. 3 ft
3. 4 ft
4. 5 ft

1-14. Resistance arm in figure 1C

1. 1 ft
2. 3 ft
3. 4 ft
4. 5 ft

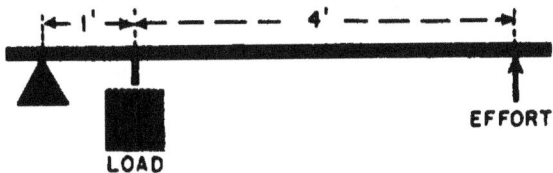

Figure 1C

1-15. Two boys find that they can balance each other on a plank if one sits six feet from the fulcrum and the other eight feet. The heavier boy weighs 120 pounds. How much does the lighter boy weigh?

1.   90 lb
2.   106 lb
3.   112 lb
4.   114 lb

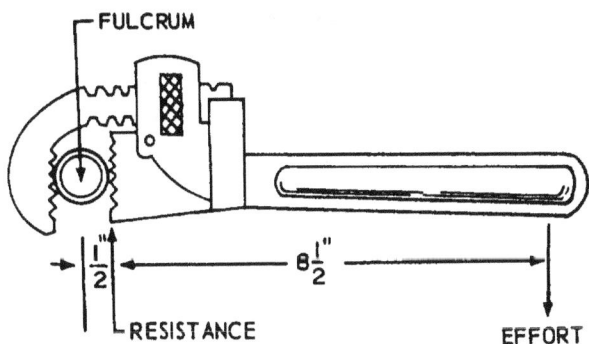

Figure 1D

1-16. With the aid of the pipe wrench shown in figure 1D, how many pounds of effort will you need to exert to overcome a resistance of 900 pounds?

1.   25 lb
2.   50 lb
3.   75 lb
4.   100 lb

Questions 1-17 and 1-18 are related to a 300-pound load of firebrick stacked on a wheelbarrow. Assume that the weight of the firebrick is centered at a point and the barrow axle is 1 1/2 feet forward of the point.

1-17. If a Seaman grips the barrow handles at a distance of three feet from the point, how many total pounds will the Seaman have to lift to move the barrow?

1.   65 lb
2.   100 lb
3.   150 lb
4.   300 lb

1-18. If a Seaman grasps the handles 3 1/2 feet from the point where the weight is centered, how many pounds of effort will be exerted?

1.   50 lb
2.   90 lb
3.   100 lb
4.   120 lb

1-19. In lever problems, the length of the effort arm multiplied by the effort is equal to the length of the

1.   resistance arm multiplied by the effort
2.   resistance arm multiplied by the resistance
3.   effort arm multiplied by the resistance arm
4.   effort arm multiplied by the resistance

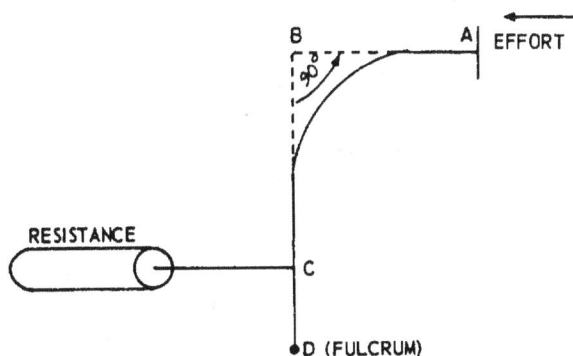

Figure 1E.—A curved lever.

1-20. The length of the effort arm in figure 1E is equal to the length of the

1.   curved line from A to C
2.   curved line from A to D
3.   straight line from B to C
4.   straight line from B to D

3

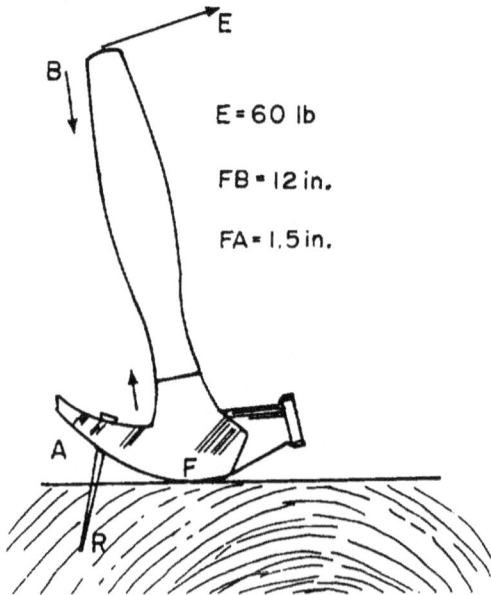

Figure 1F

E = 60 lb

FB = 12 in.

FA = 1.5 in.

1-21. Refer to figure 1F. If a person exerts at point B a pull of 60 pounds on the claw hammer shown, what is the resistance that the nail offers?

1. 60 lb
2. 120 lb
3. 480 lb
4. 730 lb

1-22. Which of the following definitions describes the mechanical advantage of the lever?

1. Effort that must be applied to overcome the resistance of an object divided by the resistance of the object
2. Amount of work obtained from the effort applied
3. Gain in power obtained by the use of the lever
4. Resistance offered by an object divided by the effort which must be applied to overcome this resistance

1-23. The mechanical advantage of levers can be determined by dividing the length of the effort arm by the

1. distance between the load and the point where effort is applied
2. distance between the fulcrum and the point where effort is applied
3. distance between the load and the fulcrum
4. amount of resistance offered by the object

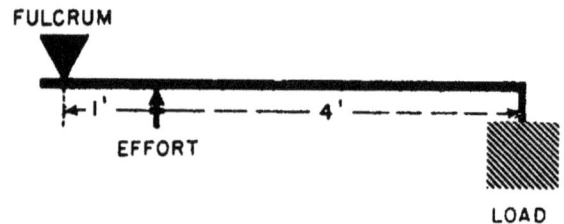

Figure 1G

1-24. The mechanical advantage of the lever in figure 1G is

1. one-fifth
2. one-fourth
3. four
4. five

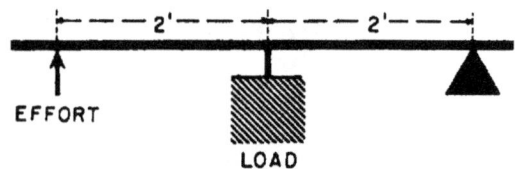

Figure 1H

1-25. The mechanical advantage of the lever in figure 1H is

1. one
2. two
3. one-half
4. one-fourth

4

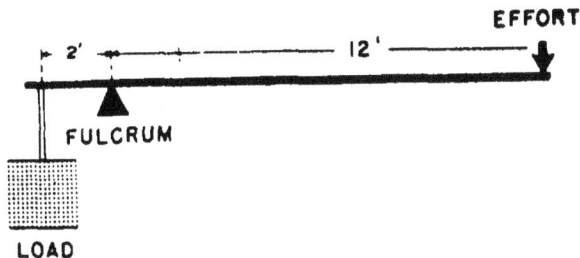

Figure 1J

1-26. The mechanical advantage of the lever pictured in figure 1J is

1. five
2. six
3. seven
4. one-sixth

1-27. The combination dog and wedge of textbook figure 1-10 is a complex machine since it consists of which two simple machines?

1. Lever and the screw
2. Two first-class levers
3. Lever and the inclined plane
4. One first-class lever and one second-class lever

Information for questions 1-28 and 1-29: The handle of a hatch dog is 9 inches long. The short arm is 3 inches long.

1-28. What is the mechanical advantage of the hatch dog?

1. 12
2. 27
3. 3
4. 9

1-29. With how much force must you push down on the handle to exert 210 pounds force on the end of the short arm?

1. 105 lb
2. 80 lb
3. 70 lb
4. 25 lb

1-30. The rope in a block and tackle is called a

1. runner
2. line
3. fall
4. sheave

1-31. The theoretical mechanical advantage of the single sheave block of textbook figure 2-2 is

1. one
2. two
3. one-half
4. zero

1-32. A single block-and-fall rigged as a runner has a theoretical mechanical advantage of

1. one
2. two
3. one-half
4. four

1-33. In a block and tackle having a mechanical advantage greater than one, how does the distance the load moves compare with the length of the rope which is pulled through the block?

1. It is less
2. It is the same
3. It is greater
4. It depends on the weight of the load

1-34. What advantage can you obtain by replacing the single fixed block of textbook figure 2-3 with the gun tackle purchase of textbook figure 2-6?

1. You can pull the rope from a more convenient position
2. You need to exert about 1/3 as much effort to lift the same load
3. You can lift the same load in 1/2 the time
4. You need to exert about 1/2 as much effort to lift the same load

5

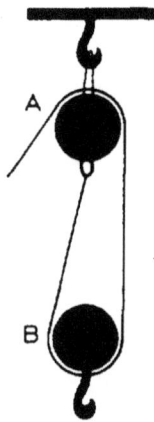

Figure 1K

1-35. In the arrangement of figure 1K the purpose of block A is to

1. increase the mechanical advantage of the block
2. change the direction of the applied force
3. hold up block B
4. act as a runner for block B

1-36. A luff tackle is a block and tackle consisting of a

1. fixed double block and a movable single block
2. movable double block and a fixed single block
3. fixed single block and a movable single block
4. fixed triple block and a movable double block

Information for questions 1-37 and 1-38: Alone you're going to hoist a 600-pound load to a height of 36 feet. You can pull 160 pounds' worth. You have to use a fixed block fastened to a beam above you. You have a movable block attached to the pad eye of the load.

1-37. What minimum mechanical advantage must the block and tackle provide?

1. One
2. Two
3. Three
4. Four

1-38. For which requirement will it be to your advantage to rig a yard to a stay tackle if each tackle has a theoretical mechanical advantage of two?

1. A theoretical mechanical advantage of 4
2. A change in the direction of pull for convenience
3. A heavy crate to be lifted to the other side of a low fence
4. An increase in speed

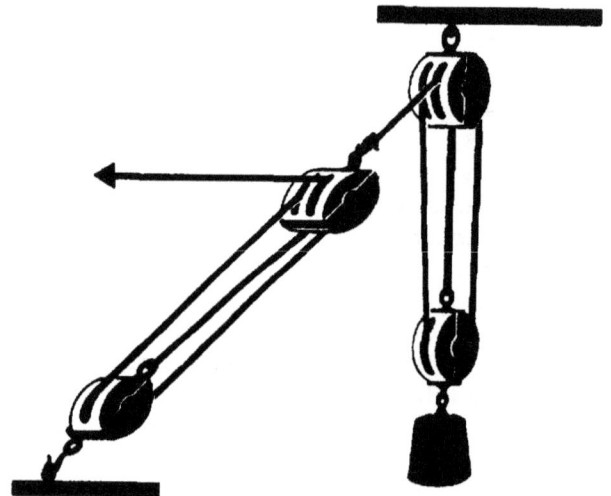

Figure 1L

1-39. The overall mechanical advantage in figure 1L is about

1. five
2. six
3. eight
4. twelve

1-40. You are using a differential pulley to lift a load of 2,400 pounds. Fifty pounds of effort are required to overcome the frictional resistance of the pulley. What force is necessary to lift the load if the theoretical mechanical advantage of the pulley is 24?

1. 50 lb
2. 100 lb
3. 150 lb
4. 200 lb

6

1-41. With a block and tackle the effort has to move 125 feet in order to raise a load 25 feet. The friction is so great that it takes a force of 75 pounds to lift a load of 300 pounds. The actual mechanical advantage is

1. five
2. two
3. three
4. four

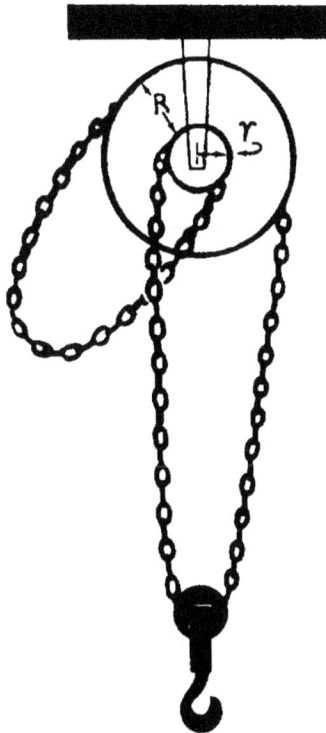

Figure 1M

1-42. The theoretical mechanical advantage of a differential pulley, such as the one pictured in figure 1M, depends upon the

1. difference in diameters of the two top pulleys
2. sum of diameters of the two top pulleys
3. length of the chain
4. difference in diameters of the two small pulleys

1-43. In the differential pulley pictured in figure 1M, if the radius of the small pulley at the top is 3 inches, the radius of the large pulley at the top is 4 inches, and the radius of the pulley at the bottom is 2 1/2 inches, the theoretical mechanical advantage is

1. 8
2. 9
3. 30
4. 36

1-44. Why is the actual mechanical advantage of the differential pulley of textbook figure 2-11 never so great as the theoretical mechanical advantage of the pulley?

1. Part of the effort applied to the chain is used to overcome the frictional resistance of the pulley's moving parts
2. The diameter of C is between those of A and B
3. The diameter of A is greater than that of B
4. The length of the chain fed down is greater than the length of the chain fed up

1-45. A wheel and axle can rotate clockwise or counterclockwise about an axis to provide a mechanical advantage or an increase in speed.

1. True
2. False

1-46. The mechanical advantage of a wheel and axle depends upon the

1. amount of force applied and the size of the wheel
2. size of the wheel and the amount of the resistance
3. ratio of the radius of the wheel to which force is applied to the radius of the axle on which it turns
4. length of the axle

1-47. What maximum load can you lift by applying a 50-pound force to the handle of an 18-inch crank that is connected to a 9-inch-diameter drum of a hand winch?

1.   50 lb
2.   100 lb
3.   150 lb
4.   200 lb

1-48. The moment resulting from a force acting on a wheel and axle is equal to the

1.   amount of force required to produce equilibrium in a wheel and axle
2.   ratio of the force to the distance from the center of rotation
3.   distance from the point where the force is applied to the center of the axle
4.   product of the amount of the force and the distance of the force from the center of rotation

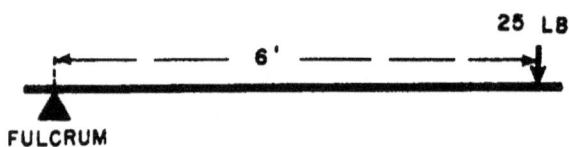

Figure 1N

1-49. The clockwise moment of force about the fulcrum of figure 1N is

1.   4 2/3 ft-lb
2.   6 ft-lb
3.   25 ft-lb
4.   150 ft-lb

1-50. If in the lever shown in figure 1N both the amount of force and the distance between the fulcrum and the point where force is applied are doubled, the torque will be

1.   1/2 as great as before the changes were made
2.   2 times as great as before the changes were made
3.   4 times as great as before the changes were made
4.   8 times as great as before the changes were made

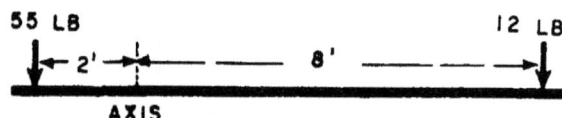

Figure 1P

1-51. What would be the resultant torque in figure 1P?

1.   Clockwise torque of 10 ft-lb
2.   Clockwise torque of 14 ft-lb
3.   Counterclockwise torque of 10 ft-lb
4.   Counterclockwise torque of 14 ft-lb

1-52. What will happen to a machine when clockwise and counterclockwise moments of force are in balance?

1.   The machine will break down
2.   The machine will either remain at rest or move at a steady speed
3.   The machine will move at an increasing speed
4.   The machine will move at a decreasing speed

8

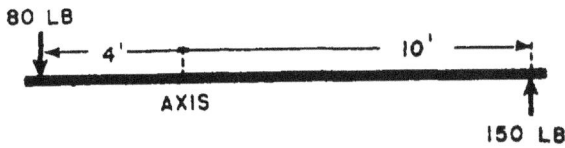

Figure 1Q

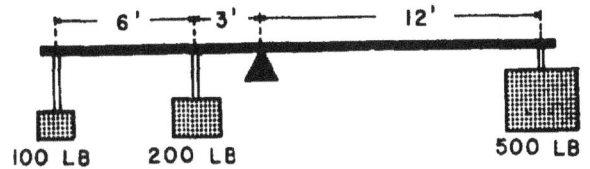

Figure 1R

1-53. The result of forces acting as shown in figure 1Q would be a torque of

1. 600 ft-lb
2. 1,180 ft-lb
3. 1,820 ft-lb
4. 2,680 ft-lb

Information to answer questions 1-54 through 1-56: The service manual for an engine states that a certain nut is to be tightened by a moment of 90 foot-pounds.

1-54. If a wrench 18 inches long is used, the amount of force that should be exerted at the end of the wrench is

1. 5 lb
2. 9 lb
3. 60 lb
4. 162 lb

1-55. How many pounds of effort could be saved by using a two-foot long wrench?

1. 15 lb
2. 30 lb
3. 45 lb
4. 50 lb

1-56. What kind of wrench could you use that measures directly the amount of force you are exerting on the nut?

1. Pipe wrench
2. Torque wrench
3. Spanner wrench
4. Adjustable end wrench

1-57. The result of forces operating as shown in figure 1R is equivalent to a moment of

1. 300 ft-lb in a clockwise direction
2. 700 ft-lb in a counterclockwise direction
3. 4,500 ft-lb in a clockwise direction
4. 6,000 ft-lb in a counterclockwise direction

When answering questions 1-58 through 1-60, refer to figure 1S.

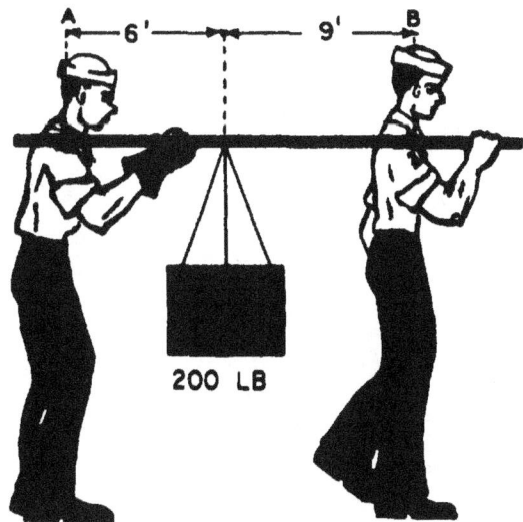

Figure 1S

1-58. The clockwise moment about A is

1. 200 ft-lb
2. 300 ft-lb
3. 1,200 ft-lb
4. 1,800 ft-lb

9

1-59. The counterclockwise moment about B is

1.   200 ft-lb
2. 1,200 ft-lb
3. 1,800 ft-lb
4. 3,000 ft-lb

1-60. How much of the load is the sailor at the right carrying?

1.   22 2/9 lb
2.   33 1/3 lb
3.   80 lb
4.   120 lb

1-61. The sailor in figure 3-4 in your textbook can increase his effectiveness without exerting a greater effort by using a shorter capstan bar.

1. True
2. False

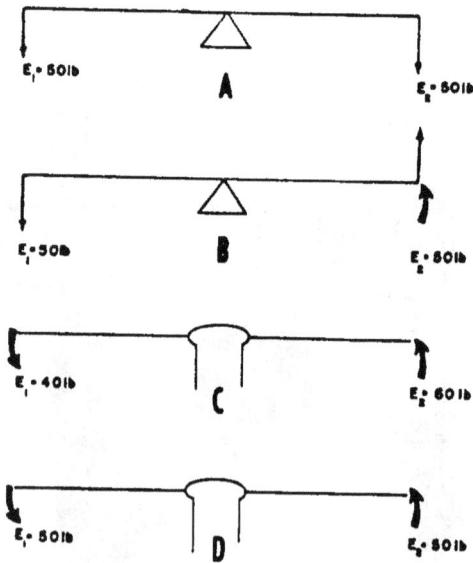

Figure 1T

1-62. Which of the parts of figure 1T represents the wheel and axle arrangement known as a couple?

1. A
2. B
3. C
4. D

1-63. A ship's deck is 24 feet above the dock. How long a gangplank is needed to provide a theoretical mechanical advantage of 2?

1. 24 ft
2. 48 ft
3. 60 ft
4. 96 ft

1-64. A sailor is rolling a 400-pound barrel up a 20-foot long ramp to a 3-foot height. Neglecting friction, the force needed to move the barrel up the ramp is

1.   60 lb
2. 133 1/3 lb
3. 200 lb
4. 220 lb

When answering questions 1-65 through 1-68, refer to figure 1U.

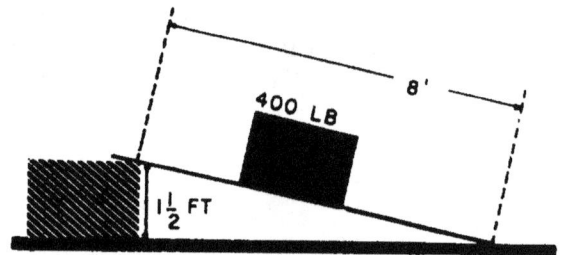

Figure 1U

1-65. The theoretical mechanical advantage of the inclined plane is

1.   3/16
2. 3
3. 5 1/3
4. 6

1-66. Neglecting friction, the force, needed to pull the crate up the inclined plane is

1.   50 lb
2.   75 lb
3. 124 lb
4. 600 lb

1-67. If a force of 133 pounds is actually required to move the crate up the inclined plane, the amount of force expended in overcoming friction is

1. 20 lb
2. 33 lb
3. 58 lb
4. 66 lb

1-68. Since a force of 133 pounds was exerted in moving the crate up the inclined plane, the actual mechanical advantage is

1. 1/3
2. 2 1/3
3. 3
4. 4

1-69. A 3,000-pound automobile is towed up a ramp 150 feet long running from the street floor to the second floor of a garage. The towing force required was 300 pounds. What is the distance between floors if 20 pounds of force were needed to overcome frictional resistance?

1. 8.6 ft
2. 10 ft
3. 14 ft
4. 15 ft

Information for questions 1-70 and 1-71: By exerting an effort of 115 pounds, you move a 300-pound crate up an inclined plane 12 feet long to a truck bed three feet above the sidewalk.

1-70. What is the theoretical mechanical advantage of the inclined plane?

1. One
2. Two
3. Three
4. Four

1-71. How much of your effort is used to overcome friction?

1. 35 lb
2. 40 lb
3. 75 lb
4. 115 lb

1-72. What is the characteristic shape of wedges which have a high mechanical advantage?

1. Short and thick
2. Long and thin
3. Long and thick
4. Short and thin

1-73. If a wedge is 6 inches long, 3 inches wide, and 1 1/2 inches thick at the top, the theoretical mechanical advantage is

1. 1 1/2
2. 2
3. 3
4. 4

1-74. A member of a damage control party uses a maul to drive a wedge in behind a shore to tighten up a damaged bulkhead. The wedge is 15 inches long and 3 inches thick at the butt. How many pounds of force will be delivered against the face of the wedge by an 80-pound blow on the wedge butt with the maul?

1. 80 lb
2. 240 lb
3. 400 lb
4. 1,200 lb

1-75. A 60-pound blow delivered against the 3/4-inch-thick butt end of a wedge results in an effective splitting force of 480 pounds. How long is the wedge?

1. 3 in
2. 4 in
3. 6 in
4. 8 in

ASSIGNMENT 2

Textbook Assignment: "The Screw," chapter 5, pages 5-1 through 5-4; "Gears," chapter 6, pages 6-1 through 6-8; "Work," chapter 7, pages 7-1 through 7-6; and "Power," chapter 8, pages 8-1 through 8-4.

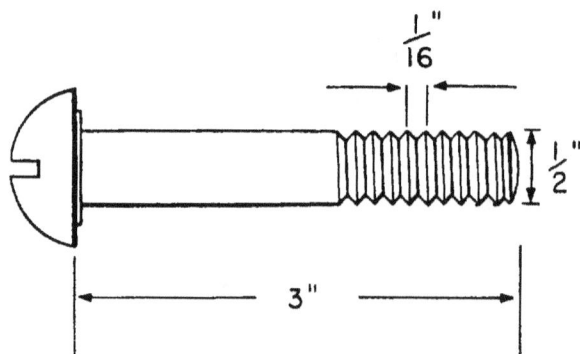

Figure 2A

2-1.   What is the pitch of the screw in figure 2A?

1.   1/16 in
2.   1/2 in
3. 1 4/7 in
4. 3   in

2-2.   Upon which measurements does the theoretical mechanical advantage of a jackscrew depend?

1.   Pitch and length of the screw
2.   Length of the jack handle and radius of the screw
3.   Pitch and radius of the screw
4.   Length of the jack handle and pitch of the screw

2-3.   How do you find the theoretical mechanical advantage of a jackscrew?

1.   Divide the amount of resistance by the amount of effort required to overcome the resistance
2.   Multiply the length of the jack handle by the radius of the screw and then divide by the length of the screw
3.   Multiply the length of the jack handle by 2π and then divide by the pitch of the screw
4.   Divide the length of the jack handle by 2π and then multiply by the pitch of the screw

2-4.   High friction losses are built into a jackscrew in order to prevent the

1.   screw from turning under the weight of a load as soon as the lifting force is removed
2.   screw from becoming overheated when a load is being lifted
3.   threads of the screw from being sheared off by the weight of a load
4.   jack from toppling over as soon as the lifting force is removed

2-5.   If a screw has a pitch of 1/16 inch, how many turns are required to advance it 1/2 inch?

1.   2
2.   8
3.   16
4.   32

2-6. If the handle of a jackscrew is turned 16 complete revolutions to raise the jack 2 inches, the pitch of the screw is

1. 1/32 in
2. 1/16 in
3. 1/8 in
4. 1/4 in

2-7. You are pulling a 21-inch lever to turn a jackscrew having a pitch of 3/16 inch. The theoretical mechanical advantage of the jackscrew is about

1. 1,000
2. 700
3. 400
4. 100

2-8. A jackscrew has a handle 35 inches long and a pitch of 7/32 inch. If a pull of 15 pounds is required at the end of the handle to lift a 3,000-pound load, the force expended in overcoming friction is

1. 12 lb
2. 9 lb
3. 3 lb
4. 5 lb

2-9. Refer to textbook figure 5-3. How many complete turns of the thimble are required to increase the opening of the micrometer by 1/4 inch?

1. 25
2. 10
3. 5
4. 4

When answering items 2-10 and 2-11, refer to textbook figure 5-4.

2-10. If the micrometer's thimble is turned exactly five complete revolutions, the new reading is

1. 0.753 in
2. 0.703 in
3. 0.628 in
4. 0.517 in

2-11. Assume that the graduation mark 5 on the thimble is opposite point X. How much farther do you open the micrometer in turning the thimble until the graduation mark 15 is opposite the point for the first time?

1. 0.125 in
2. 0.010 in
3. 0.0125 in
4. 0.0010 in

2-12. How do you find the actual mechanical advantage that a jackscrew provides in lifting a load?

1. Multiply the length of the jack handle by the radius of the screw and then divide by the pitch of the screw
2. Divide the load by the amount of effort required to lift the load
3. Multiply the length of the jack handle by 2 and then divide by the pitch of the screw
4. Divide the distance the screw travels by the number of turns it makes and then subtract the amount of frictional resistance

2-13. If a jackscrew has a pitch of 5/32 inch, the length of the handle required to obtain a theoretical mechanical advantage of 800 is about

1. 30 in
2. 25 in
3. 20 in
4. 15 in

2-14. If a jackscrew requires a force of 15 pounds at the end of the handle to lift a 3,000 pound load, its actual mechanical advantage is

1. 4,500
2. 2,000
3. 450
4. 200

2-15. Which of the following describes the cut of the threads in a screw gear?

1. One end has left-hand threads and the other has right-hand threads
2. Both ends have left-hand threads
3. Both ends have right-hand threads
4. One end has a greater pitch and less depth than the other

2-16. Two Seamen are using a quadrant davit to put a large lifeboat over the side. If the operating handle is released while the boat is being lowered, the boat is kept from falling by means of

1. a friction brake on the operating handle
2. a davit arm and swivel
3. a counterweight
4. self-locking threads on the screw

2-17. Gears serve all of the following purposes EXCEPT

1. eliminating frictional losses
2. changing the direction of motion
3. increasing or decreasing the applied force
4. increasing or decreasing the speed of the applied motion

2-18. What condition must hold true if two gears are to mesh properly?

1. The teeth of both gears must be the same size
2. Both gears must have the same diameter
3. The teeth must be cut slanting across the working faces of the gears
4. The gears must turn on parallel shafts

2-19. Herringbone gears are sometimes used instead of single helical gears in order to

1. change the direction of motion
2. increase the mechanical advantage
3. increase the gear ratio
4. prevent axial thrust on the shaft

2-20. If you should find it necessary to transmit circular motion from a shaft to a second shaft, which is at right angles to the first shaft, which of the following gear arrangements should you use?

1. Internal and pinion gears
2. Miter gears
3. Spur gears and idler
4. Rack and pinion gears

2-21. In a worm and spur gear arrangement, the worm gear is single-threaded and has six threads, and the spur gear has 30 teeth. In order to turn the spur gear one complete revolution, the worm gear must be given how many complete turns?

1. 5
2. 30
3. 50
4. 180

2-22. If the worm gear in the worm and spur gear arrangement in question 2-21 were triple-threaded, the number of times the worm gear would have to be turned in order to produce one complete revolution of the spur gear would be

1. 3 times
2. 10 times
3. 15 times
4. 60 times

2-23. You have a pinion gear with 14 teeth driving a spur gear with 42 teeth. If the pinion turns at 420 rpm, what will be the speed of the spur gear?

1. 42 rpm
2. 140 rpm
3. 160 rpm
4. 278 rpm

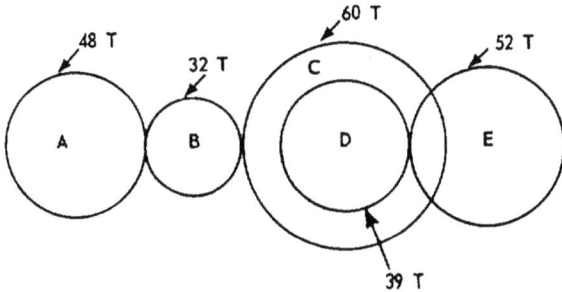

Figure 2B

For items 2-24 through 2-28, refer to the gear system in figure 2B and to the symbols which follow.

Gears C and D are rigidly attached to one another.

S_a = speed of gear A
S_b = speed of gear B
S_c = speed of gear C
S_d = speed of gear D
S_e = speed of gear E
A = number of teeth on gear A
B = number of teeth on gear A
C = number of teeth on gear C
D = number of teeth on gear D
E = number of teeth on gear E

2-24. Given S_a, A, and B, a formula for testing S_b is

1. $S_a/S_b = A/B$
2. $S_b/S_a = A/B$
3. $S_a S_b = AB$
4. $S_b = B/A - S_a$

2-25. Given S_a, A, B, C, D, and E, a formula for finding S_e is

1. $S_e = S_a \dfrac{(ABD)}{(BCE)}$

2. $S_e = S_a \dfrac{(ABE)}{(CDE)}$

3. $S_e/S_a = E/A$

2-26. Gear A would make how many revolutions for every complete revolution of gear C?

1. 4/5
2. 1 1/5
3. 1 1/4
4. 1 1/2

2-27. Which formula is used for finding the mechanical advantage of the system of gears including only gears B, C, D, and E, assuming that power is applied to gear B?

1. $\dfrac{C}{B} \times \dfrac{E}{D}$

2. $\dfrac{B}{C} \times \dfrac{D}{E}$

3. $\dfrac{C}{B} + \dfrac{E}{D}$

4. $\dfrac{B}{C} + \dfrac{D}{E}$

2-28. Assuming that power is applied to gear A, the entire gear train will have a mechanical advantage of

1. 0.67
2. 1.67
3. 3.34
4. 6.68

15

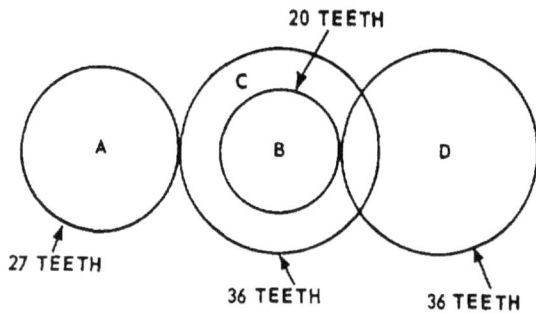

Figure 2C

2-29. Gears B and C in the gear arrangement shown in figure 2C are rigidly fixed together. If gear A is turned counterclockwise at a rate of 120 rpm, in what direction and at what rate will gear D turn?

1. Clockwise at 20 rpm
2. Clockwise at 50 rpm
3. counterclockwise at 50 rpm
4. Counterclockwise at 100 rpm

2-30. The product of all the driving teeth of a turbine reduction gearing is 400 and the product of the driven teeth is 4,000. When the output shaft turns at 200 rpm, the turbine turns at

1. 200 rpm
2. 400 rpm
3. 2,000 rpm
4. 4,000 rpm

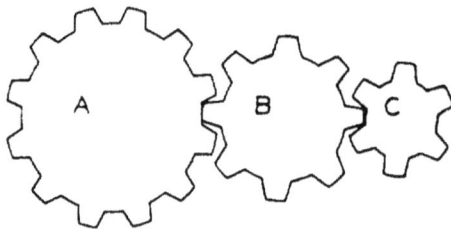

Figure 2D

2-31. The speed ratio of the gear train in figure 2D is 2 to 1. If gear B is removed and gear C is placed so that it runs directly off gear A. the speed ratio will be

1. 2.0 to 1
2. 2.4 to 1
3. 3.0 to 1
4. 4.0 to 1

2-32. The purpose of an idler gear is to

1. increase the speed ratio
2. take up lost motion
3. change the direction of rotation
4. keep another gear in place

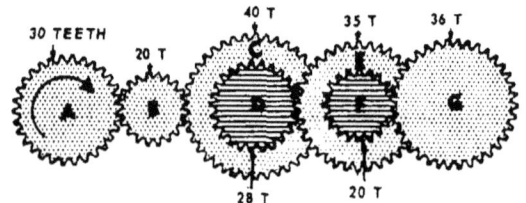

Figure 2E

Items 2-33 through 2-36 are based on the gear train shown in figure 2E.

2-33. Which gear serves as an idler gear?

1. D
2. C
3. B
4. A

2-34. If gear A turns at 300 rpm, how fast does gear G turn?

1. 180 rpm
2. 100 rpm
3. 87 rpm
4. 80 rpm

2-35. What is the mechanical advantage of the train?

1. Five
2. Two
3. Three
4. Four

16

2-36. The direction of rotation of gear G is counterclockwise.

1. True
2. False

Use the following information to compute the mechanical advantage of textbook figure 6-1: Gear A radius = 2 inches; Gear A teeth = 36; Gear B and C teeth = 8; handle turn radius = 1 1/2 inches.

2-37. The mechanical advantage of the eggbeater is

1. 1/8
2. 1/6
3. 1/4
4. 1/2

2-38. Refer to textbook figure 6-3B. What is the function of this gear arrangement if the pinion is driving the internal gear?

1. To increase speed
2. To magnify force
3. To change direction of motion
4. To change rotary motion into linear motion

2-39. Refer to the left-hand half of figure 6-12 in your textbook. Which of the following statements best describes the action of the valve as the camshaft rotates 180° from its position as shown in the figure?

1. The valve remains closed
2. The valve opens and stays open
3. The valve opens and then closes
4. The valve opens and closes twice

2-40. A foot-pound is defined as the amount of

1. force developed by a one-pound weight falling a distance of one foot
2. energy required to lift a one-pound weight
3. power required to overcome a resistance of one pound
4. work required to overcome a resistance of one pound through a distance of one foot

2-41. Which of the following is an example of work?

1. Holding two pieces of glued wood in a vise
2. Rolling a barrel up a gangplank
3. Changing water to steam
4. Burning a log in a fireplace

2-42. When you calculate the amount of work you have done on an object, the factors which you must always measure are the

1. resistance encountered and the distance it is moved
2. weight of the object and the distance it is moved
3. angle at which force is applied and weight of the object
4. time required to move the object and resistance encountered

2-43. Assume that you must apply a force of 150 pounds to overcome the resistance of a crate weighing 350 pounds. In moving the crate up an inclined plane which is 12 feet long, how much work do you do?

1. 4,200 ft-lb
2. 1,800 ft-lb
3. 350 ft-lb
4. 150 ft-lb

Information for items 2-44 through 2-46: You are using a first-class lever to raise a 400-pound load to a height of 1 foot. The effort arm of your lever is 8 feet long and the resistance arm is 2 feet long.

2-44.   How much work is done in raising the load?

    1.   400 ft-lb
    2.   300 ft-lb
    3.   200 ft-lb
    4.   50 ft-lb

2-45.   How far must you move the lever in order to raise the load 1 foot?

    1.   1 ft
    2.   2 ft
    3.   8 ft
    4.   4 ft

2-46.   How much work is done in balancing the load at the 1-foot height?

    1.   0 ft-lb
    2.   2 ft-lb
    3.   8 ft-lb
    4.   10 ft-lb

2-47.   By using a machine to move an object you can

    1.   decrease the amount of work to be done
    2.   reduce the weight of the object
    3.   decrease the amount of the effort required
    4.   reduce the resistance of the object

Information for questions 2-48 through 2-51: You are using a 24,000-pound load with a screwjack that has a pitch of 1/4 inch and a 24-inch handle.

2-48.   Theoretically, (by neglecting friction), you should be able to turn the jack handle by exerting an effort of about

    1.   24 lb
    2.   40 lb
    3.   60 lb
    4.   80 lb

2-49.   Because of friction. you actually have to apply a 120-pound force to turn the jack handle. About how much work do you do in turning the handle one complete revolution?

    1.   120 ft-lb
    2.   240 ft-lb
    3.   1,500 ft-lb
    4.   3,000 ft-lb

2-50.   With each revolution of the jack handle, the work output of the jack equals

    1.   100 ft-lb
    2.   200 ft-lb
    3.   500 ft-lb
    4.   600 ft-lb

2-51.   The efficiency of the jack is

    1.   12 1/2%
    2.   33 1/3%
    3.   50   %
    4.   66 2/3%

Information for questions 2-52 through 2-54: You push with a force of 125 pounds to slide a 250-pound crate up a gangplank. The gangplank is 12 feet long and the upper end is 5 feet above the lower end.

2-52.   What is the theoretical mechanical advantage of the gangplank?

    1.   1
    2.   2
    3.   2.4
    4.   12

2-53.   How much of your 125-pound push is used to overcome friction?

    1.   21   lb
    2.   42   lb
    3.   62 1/2 lb
    4. 125   lb

2-54.   What is the efficiency of the gangplank?

    1.   25%
    2.   50%
    3.   75%
    4.   83.3%

Information for questions 2-55 and 2-56: You want to raise an 1,800-pound motor 4 feet up to a foundation. You use two double-sheave blocks rigged to give a mechanical advantage of 4 and a windlass that has a theoretical mechanical advantage of 6.

2-55. Assuming 100 percent efficiency, how much work is required to raise the motor?

1. 1,800 ft-lb
2. 3,600 ft-lb
3. 7,200 ft-lb
4. 10,800 ft-lb

2-56. Neglecting friction, how much pull must you exert to raise the motor?

1. 18 lb
2. 36 lb
3. 75 lb
4. 300 lb

2-57. An effect which friction has on the mechanical advantage of any machine is to make the

1. theoretical mechanical advantage less than the actual mechanical advantage
2. actual mechanical advantage less than the theoretical mechanical advantage
3. actual mechanical advantage less than one
4. actual mechanical advantage more than one

2-58. Assume that the hammer of a pile driver weighs 1,000 pounds. The resistance of the earth is 6,000 pounds. If the hammer drops 4 feet to drive a pile, how far into the earth will the pile be driven? (Assume an efficiency of 100%.)

1. 2 in
2. 6 in
3. 8 in
4. 10 in

When answering questions 2-59 through 2-61, assume that a man lifts a 600-pound load, using a block and tackle with a theoretical mechanical advantage of 6. He does 6,500 foot-pounds of work in lifting the load 8 feet.

2-59. How much work does the man do in overcoming friction?

1. 215 ft-lb
2. 813 ft-lb
3. 1,700 ft-lb
4. 5,900 ft-lb

2-60. The total force exerted by the man in lifting the load is approximately

1. 35 lb
2. 135 lb
3. 215 lb
4. 406 lb

2-61. The average amount of force which the man exerted to overcome friction is approximately

1. 35 lb
2. 215 lb
3. 237 lb
4. 406 lb

2-62. The handle of a screwjack must move through a circular distance of 600 inches to lift a load one inch. If a force of 10 pounds is required to lift a load of 1,500 pounds, what is the efficiency of the jack?

1. 25%
2. 33%
3. 78%
4. 90%

2-63. A block and tackle has a theoretical mechanical advantage of 4 but requires a force of 50 pounds to lift a 160-pound load. The efficiency of the block and tackle is

1. 60%
2. 70%
3. 80%
4. 90%

2-64. In a certain machine, the effort moves 20 feet for every foot that the resistance moves. If the machine is 75 percent efficient, the force required to overcome a resistance of 300 pounds is

1. 15 lb
2. 20 lb
3. 25 lb
4. 30 lb

2-65. If a block and tackle has a theoretical mechanical advantage of 5 and an efficiency of 60 percent, the amount of force necessary to lift a 1,200-pound load is

1. 30 lb
2. 150 lb
3. 400 lb
4. 720 lb

2-66. Which of the following statements concerning the relationship of work output and work input of a machine is correct?

1. The output is the same as the input
2. The output is greater than the input
3. The output is less than the input
4. The output has no relationship to the input

2-67. The amount of work done divided by the time required is called

1. energy
2. resistance
3. force
4. power

When answering questions 2-68 through 2-73, assume 100 percent efficiency in each situation and use the appropriate power formula to calculate the unknown quantity.

2-68. A motor-driven hoist lifts a 165-pound load to a height of 50 feet in 30 seconds. How much power does the motor develop?

1. 1/4 hp
2. 1/2 hp
3. 3 hp
4. 10 hp

2-69. A power winch is capable of lifting a 440-pound load a distance of 5 feet in 1 second. The driving motor works at the rate of

1. 1/2 hp
2. 1 hp
3. 2 hp
4. 4 hp

2-70. What is the horsepower of the engine driving the pump that lifts 9,900,000 pounds of water per day from a lake to the top of a standpipe, a vertical distance of 120 feet? The engine runs at a uniform speed 12 hours a day.

1. 12 hp
2. 15 hp
3. 24 hp
4. 50 hp

2-71. While a propeller-driven aircraft travels at a speed of 120 mph, its engine develops 1,500 hp. Approximately what force in pounds is being exerted by the propeller?

1. 850 lb
2. 5,000 lb
3. 15,000 lb
4. 30,000 lb

2-72. What is the horsepower of a hoisting engine that can raise 6,000 pounds through a height of 44 feet in one minute?

1. 3 hp
2. 4 hp
3. 8 hp
4. 12 hp

2-73. An annunition hoist is powered by a 2-hp motor. Working at full load, how long does it take the motor to raise a 50-pound shell 22 feet from the handling room to the gun turret?

1. 1/2 sec
2. 1 sec
3. 1 1/2 sec
4. 2 sec

2-74. If it is desired to develop ten usable horsepower from an engine which is 50 percent efficient, the engine must have a minimum rated horsepower of at least

1. 10
2. 20
3. 100
4. 150

2-75. What information is sufficient to find the horsepower rating of a motor by means of the Prony brake in figure 8-3 of the textbook?

1. The readings on both scales, the radius of the pulley, and the time it takes the motor to reach maximum speed
2. Tile readings on both scales. the radius of the pulley, and the speed of the motor
3. The readings on both scales, the radius of the pulley, and the diameter of the motor shaft.
4. The radius of the pulley and the readings on the scales when the belt is pulled tight enough to prevent the motor from turning

ASSIGNMENT 3

Textbook
Assignment:

"Force and Pressure," chapter 9, pages 9-1 through 9-7; "Hydrostatic and Hydraulic Machines," chapter 10, pages 10-1 through 10-10; and "Machine Elements and Basic Mechanisms," chapter 11, pages 11-1 through 11-15.

3-1. With which of the following devices is force measured?

1. A manometer
2. A bourdon gauge
3. A spring scale
4. A barometer

3-2. Pressure is expressed in terms of

1. distance and density
2. volume and force
3. density and volume
4. area and force

3-3. If a cylindrical tank which stands on end is 4 feet in diameter and contains 350 pounds of water, the pressure on the bottom of the tank is approximately

1. 22 lb per sq ft
2. 28 lb per sq ft
3. 65 lb per sq ft
4. 350 lb per sq ft

3-4. At sea level, what is the force of the atmosphere on each side of a cube measuring 16 inches on a side?

1. 380 lb
2. 890 lb
3. 2,400 lb
4. 3,840 lb

3-5. If the pressure in a steam boiler that supplies pressure to a piston 4 inches in diameter is 600 pounds per square inch, the total force exerted on the piston is approximately

1. 150 lb
2. 600 lb
3. 2,400 lb
4. 7,500 lb

3-6. The airbrake cylinder on a railroad car has a diameter of 8 inches. The locomotive supplies compressed air to this cylinder at 90 pounds pressure per square inch. How much force is transmitted to the brake shoes when the brakes are applied?

1. 720 lb
2. 4,520 lb
3. 5,000 lb
4. 6,500 lb

3-7. When the pressure being measured with the gauge shown in textbook figure 9-4 is decreased, the linkage end of the Bourdon tube has a tendency to move so as to cause the

1. tube to become less curved
2. tube to become more curved
3. pointer to turn clockwise
4. pointer and gear to turn in opposite directions

3-8. In which of the following situations would a Schrader gauge be used instead of a Bourdon gauge or diaphragm gauge?

1. Measuring the force that air exerts on an object at sea level
2. Measuring pressure in a hydraulic system in which the load fluctuates rapidly
3. Measuring the force that water exerts on an object at the bottom of a tank
4. Measuring air pressure in the space between inner and outer boiler casings

3-9. What instrument is best for measuring pressure differences in an atmosphere of air where the pressure ranges between 31 and 32 inches of mercury?

1. Bourdon gauge
2. Schrader gauge
3. Manometer
4. Diaphragm gauge

3-10. A barometer is used to measure

1. absolute temperature
2. atmospheric pressure
3. relative humidity
4. steam pressure

3-11. The forces in an aneroid barometer that balance each other are the

1. resistance of a metal box to stretching or compression plus the tension in a spring and atmospheric pressure
2. resistance of a metal box to stretching or compression plus the force exerted by the air in the box, and the force exerted by the atmosphere
3. force exerted by steam under pressure and the tension in a spring
4. force resulting from expansion in a metal bar and the tension in a spring

3-12. The forces in a mercurial barometer which balance each other are the

1. weight of a column of mercury plus the force exerted by the air in the tube above the mercury, and the force exerted by the atmosphere plus 14.7 psi
2. force exerted by steam under pressure and the force exerted by the weight of a column of mercury
3. forces exerted by the atmosphere and the weight of a column of mercury
4. weight of a column of mercury, and the pressure of the vacuum above the mercury plus the force exerted by the atmosphere

3-13. If an airtight container is filled with steam and then cooled so that the steam condenses, the pressure inside the container is reduced because

1. a volume of steam weighs less than an equal volume of water
2. the pressure on the surface of a liquid is always zero
3. the water resulting from the condensation of the steam cannot be compressed
4. a partial vacuum results from the condensation of the steam

3-14. Which of the following instruments is used for measuring pressures in the condenser for a steam turbine?

1. Barometer
2. Schrader gauge
3. Manometer
4. Bourdon tube gauge

Information for items 3-15 and 3-16: Pressure measurements are generally classified as absolute pressure or gauge pressure. Absolute pressure is the total pressure, including that of the atmosphere; it is the pressure measured above zero pressure as a reference level. Gauge pressure is the difference between absolute pressure and the pressure of the atmosphere; it is pressure measured above atmospheric pressure as a reference level.

3-15. At sea level, the pressure in a tire is 24 psi gauge pressure. The absolute pressure in the tire is approximately

1. 39 psi
2. 33 psi
3. 24 psi
4. 9 psi

3-16. At sea level, the pressure in an air tank as measured by an aneroid barometer is 31 inches of mercury. How much greater or less than atmospheric pressure is the pressure in the air tank?

1. 1 inch more
2. 1 inch less
3. 2 inches less
4. 2 inches more

23

3-17. A manometer is an example of forces in equilibrium. The forces that balance each other are the

1. force exerted by the atmosphere and the force exerted by the liquid inside the closed container
2. force exerted by the gas inside the closed container plus the weight of the liquid on one side of the tube, and the force exerted by the atmosphere plus the weight of the liquid in the other side of the tube
3. force exerted by the steam in the steam line and the weight of a column of liquid
4. force exerted by the gas inside the closed container and the weight of part of the liquid in the tube

3-18. What instruments are interchangeable as pressure-measuring devices?

1. Aneroid barometer and mercurial barometer
2. Schrader gauge and manometer
3. Spring scale and steel yard
4. Bourdon gauge and diaphragm-type pressure gauge

3-19. Hydrostatic pressure is the pressure exerted by

1. gas in motion
2. gas at rest
3. liquid at rest
4. liquid in motion

3-20. Density is defined in terms of

1. pressure and volume
2. weight and distance
3. pressure and area
4. weight and volume

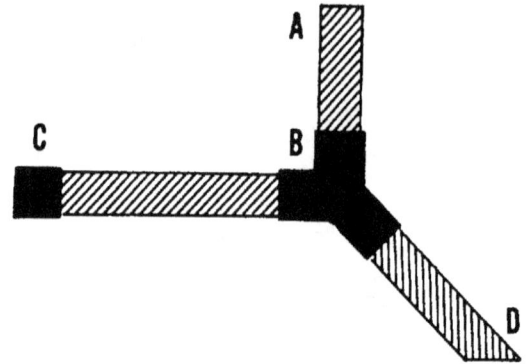

Figure 3A

3-21. The pipes in figure 3A are filled with water. Pipe AB is vertical; pipe CB is horizontal; pipe ED points downward at an angle. The point of greatest pressure is at point

1. A
2. B
3. C
4. D

3-22. The density of lead is approximately how many times greater than the density of water?

1. 5
2. 7
3. 9
4. 11

3-23. Which of the following is a true statement concerning the pressure of water on a submerged submarine?

1. The pressure is equal on the top and on the bottom
2. The pressure is greater on the top than on the bottom
3. The pressure is greater on the bottom than on the top
4. There is pressure only on the top

3-24. If one cubic foot of substance A weighs more than one cubic foot of substance B, what is the relationship between the densities of substances A and B?

1. The density of substance A is greater than the density of substance B
2. The density of substance A is less than the density of substance B
3. The density of substance A is the same as the density of substance B
4. Not enough information is given to determine the relationship

3-25. Depth charges are dropped in the vicinity of a submerged submarine. The depth charge illustrated in textbook figure 10-1 is set so as to be exploded by the

1. speed of the depth charge as it nears the submarine
2. speed of the depth charge as it enters the water
3. impact of the depth charge against the hull of the submarine
4. pressure of the water at the estimated depth of the submarine

3-26. Hydrostatic pressure in a torpedo is employed to

1. maintain the torpedo on course
2. launch the torpedo
3. keep the torpedo at desired depth
4. increase the torpedo speed

3-27. In a torpedo depth engine, the setting of the depth screw determines the

1. pressure of the air supplied to the depth engine
2. length of the pendulum
3. angular set of the vertical rudders
4. amount of force which is required to move the diaphragm

3-28. The air pumped into a diver's suit helps him or her to withstand the pressure of the water because

1. pressure of the air in the diver's suit is greater than the pressure of the water
2. air enters the diver's body so that the pressure inside his or her body is equal to the water pressure
3. air is compressible and water is not
4. force is not transmitted by air

3-29. The pressure in a diver's suit must be released gradually because

1. if pressure is released too rapidly, the air which entered the diver's body under high pressure will cause bubbles to form in his or her blood stream
2. the diver's lungs cannot quickly become adjusted to breathing air at normal pressure
3. the diver's blood circulation was partly cut off while under high pressure, and sudden return of normal circulation is painful
4. air at normal pressure contains less oxygen than air at high pressure, and the body must adjust to this condition gradually

3-30. The pitometer log determines the speed of a ship by measuring the difference between

1. hydrostatic pressure near the keel of the ship and hydrostatic pressure near the water line
2. hydrostatic pressure and the pressure of water in motion past the ship at the same depth
3. pressure of the water moving past the ship and atmospheric pressure
4. pressure of the water moving past the ship and the speed of surface wind

3-31. The speed of a ship can be determined from a pitometer log by

1. multiplying the reading on the pitometer log by a constant factor which is dependent upon the characteristics of the ship
2. combining the reading of the pitometer log with the reading of the engine revolution counter
3. dividing the reading by the density of the water
4. reading it directly from the calibrated scale

3-32. A hydraulic machine is one which operates as a result of forces transmitted by

1. mechanical energy
2. electrical energy
3. steam in a closed space
4. liquid in a closed space

3-33. A closed hydraulic system will not operate properly if air is present in the lines or cylinders because

1. air is highly compressible and cannot be used to transmit forces
2. air, being compressible, would not transmit the applied pressure
3. air interferes with the proper operation of the valves
4. air increases the pressure in both cylinders

3-34. Which of the following is NOT a true statement concerning transmission of pressure in a liquid in a closed space?

1. Pressure applied to any part of the liquid is transmitted equally to all points in the liquid
2. Pressure applied to any part of the liquid is transmitted to all points in the liquid without loss
3. Pressure in the liquid causes it to expand and increase in density
4. Pressure in the liquid acts at right angles to the walls of the container regardless of the shape of the container

3-35. Which of the following has NO relationship to the mechanical advantage of a hydraulic machine with one small and one large piston?

1. The area of the small piston
2. The area of the large piston
3. The length of the connecting tube
4. The distances the two pistons move

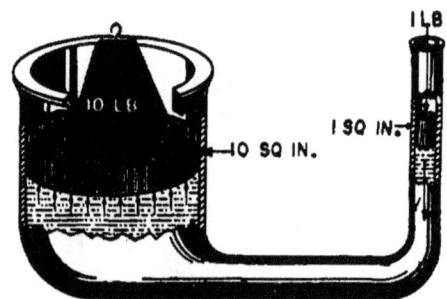

Figure 3B

26

Items 3-36 through 3-38 are related to figure 3B.

3-36. If the weight on the large piston just balances the weight on the small piston, it follows that the

1. force per unit of area is the same on both pistons
2. weights on the two pistons are equal
3. force on the large piston equals that on the small piston
4. pressure is greater below the small piston than it is below the large piston

3-37. If a certain force is applied to the small piston, what are the relationships between pressures in various parts of the system?

1. The pressure on the small piston is greater than the pressure on the large piston
2. The pressure on the small cylinder is the same as the pressure acting against the small piston and is greater than the pressure in the large cylinder
3. The pressure in the connecting tube is the same as the pressure in the small cylinder and is greater than the pressure in the large cylinder
4. The pressure is the same on all parts of all surfaces that enclose the liquid

3-38. Let F_1 be the force applied to the small piston and F_2 be the force exerted by the large piston. Which equation represents the relationship between the forces F_1 and F_2?

1. $F_2 - F_1 = 10$
2. $10 \ F_1 = F_2$
3. $F_1 + F_2 = 10$
4. $F_1 = 10 \ F_2$

3-39. The area of the small piston in a hydraulic press is 3 square inches and the area of the large piston is 75 square inches. If a force of 50 pounds is applied to the small piston, the large piston will (neglecting frictional losses) exert a force of

1.   25 lb
2.   250 lb
3.   725 lb
4.   1,250 lb

3-40. In a hydraulic press, how does the distance the small piston moves compare with the distance the large piston moves?

1. The small piston will always move a greater distance than the large piston
2. The large piston will always move a grater distance than the small piston
3. Both pistons will move the same distance
4. There is no relationship between the movements of the two pistons

Items 3-41 and 3-42 are related to the hydraulic press shown in textbook figure 10-10.

3-41. What is the main function of the check valves?

1. To prevent the liquid from escaping from the large cylinder into the reservoir
2. To prevent the liquid in the reservoir from flooding the small cylinder
3. To make possible several short strokes, instead of one long stroke, with the small piston
4. To allow the large piston to return to its starting position

3-42. What is the principle function of the globe valve?

1. To protect the cylinder from excessive pressure
2. To prevent the liquid in the reservoir from flooding the small cylinder
3. To make possible several short strokes instead of one long stroke with the piston
4. To allow the fluid in the large cylinder to flow back into the reservoir

3-43. A main ballast tank on a submarine is filled with sea water by

1. allowing air to escape from the vents at the top of the tank and allowing water to enter through flood ports at the bottom of the tank
2. pumping air from the tank and pumping water into the tank through the vents at the top
3. pumping it in through the vents at the top of the tank
4. pumping it in through the ports at the bottom of the tank

3-44. How is the water removed from the main ballast tanks when a submerged submarine is surfacing?

1. Motor-driven pumps syphon off the water
2. The water is forced out with high-pressure air
3. The water flows out through ports under the pull of gravity
4. Hydraulic pumps syphon off the water

3-45. The variable ballast tanks on a submarine are filled with sea water by

1. allowing air to escape from the tanks and water to enter through flood ports at the bottom of the tanks
2. pumping air from the tanks and allowing the water to enter through vents at the top of the tanks
3. either of the above methods
4. pumping it in

3-46. Hydraulic machines are used aboard submarines for

1. opening and closing the vent valves of the main ballast tanks
2. raising and lowering the periscope
3. opening and closing the vent valves of the safety tanks
4. all of the above purposes

When answering items 3-47 and 3-48, refer to figure 10-14 of your textbook.

3-47. The purpose of an accumulator in the hydraulic system is to

1. accumulate oil as it is released from the reservoir
2. keep the air in the system at a constant pressure
3. accumulate excess oil which flows past check valves in the system
4. keep the oil in the system under pressure

3-48. To what part, if any, is the piston in the accumulator fastened?

1. A rod which is operated by a crankshaft
2. A rod which is activated by pressurized oil in the reservoir
3. A main flood valve
4. None

3-49. Why is it easier to push a 50-pound barrel up a gangplank than to push a 50 pound box?

1. Rolling friction is less than a sliding friction
2. The shape of a barrel defies gravity better than the shape of a box
3. The barrel has a greater surface to come in contact with the gangplank
4. All of the above reasons

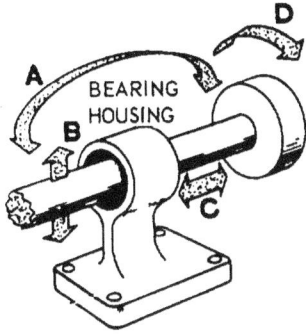

Figure 3C

Items 3-50 through 3-52 are based on figure 3C.

3-50. Which of the following types of bearing is often used in the housing to provide free movement in the direction indicated by arrow C?

1. Thrust bearing
2. Journal bearing
3. Reciprocal motion bearing
4. Tapered roller bearing

3-51. A radial ball bearing used in the housing is superior to a journal bearing for

1. reduction of friction under heavy twisting stress as indicated by arrow A
2. absorption of stress as indicated by arrow B
3. prevention of shaft motion as indicated by arrow C
4. reduction of friction during high-speed rotation of the shaft as indicated by arrow D

3-52. What type of bearing is designed to permit free rotation of the shaft while restraining motion in the direction indicated by arrow C?

1. Radial ball bearing
2. Needle roller bearing
3. Thrust bearing
4. Journal bearing

3-53. The two hardened steel rings of a ball bearing assembly are called the

1. rollers
2. races
3. separators
4. shoulders

Figure 3D

3-54. The spring in the mechanism shown in figure 3D is used to

1. store energy for part of a functioning cycle
2. force a component to engauge another component
3. return a component to neutral position after displacement
4. counterbalance a weight or thrust

29

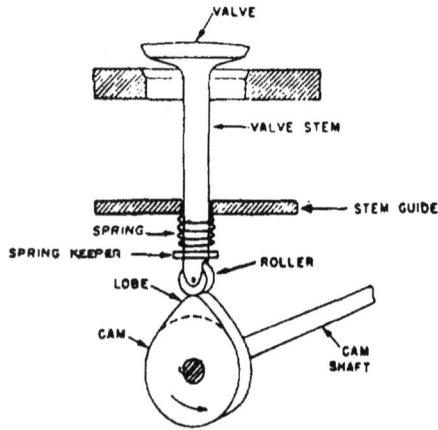

Figure 3E

3-55. The function of the spring in figure 3E is to

1. store energy for part of a cycle
2. counterbalance a weight or a thrust
3. return a component to its original position after displacement
4. permit some freedom of movement between aligned components without disengaging them

3-56. Which of the following types of springs can be used in compression, extension, or torsion?

1. Flat spring
2. Spiral spring
3. Helical spring
4. Each of the above

3-57. What are volute springs?

1. Spiral springs made of plaited strands of cable
2. Helical, conical springs wound with each coil partly overlapping the coil next to it
3. Flat springs made of slightly curved plates
4. Double cone springs with their large ends joined together

3-58. As used in some automotive suspension systems, straight torsion bars reduce shock or impact by

1. compressing
2. twisting
3. bending
4. telescoping

3-59. What gear of the gear differential is fastened to the spider shaft?

1. Input gear
2. End gear
3. Output gear
4. Spider gear

3-60. In the gear differential shown in figure 11-11 of your textbook, in proportion to the sum of revolutions of the end gears, how many revolutions does the spider make?

1. One half as many
2. The same number
3. Twice as many
4. Four times as many

3-61. Which of the following statements is true of a gear differential no matter which type of hook-up is used?

1. The spider will follow the end gears for half the sum or difference of their revolutions
2. The two side gears are the inputs and the gear on the spider shaft is the output
3. The spider shaft is one input, and one of the sides is the other output
4. If the two inputs are equal and opposite, the spider- shaft will move in either direction

3-62. Slightly worn linkages can probably be adjusted by lengthening or shortening the rods and shafts.

1. True
2. False

3-63. Rocker arms are a variation of which of the following parts?

1. The clevis
2. The lever
3. The turn buckle
4. The coupling

3-64. The counterbalance weights on the clamps of a sleeve coupling serve to

1. increase speed
2. decrease shaft vibration
3. transmit motion from a link moving in one direction to a link moving in a different direction
4. change rotary motion to linear motion

3-65. The coil spring in an Oldham coupling serves to

1. reduce friction between the coupling disks
2. keep the coupling disks in place
3. make allowance for changes in shaft length
4. strengthen the coupling

3-66. What device is used to couple two shafts that meet at a 15° angle?

1. Sleeve coupling
2. Hooke joint
3. Oldham coupling
4. Flexible coupling

3-67. The amount of whip in shafts coupled by a Hooke joint depends on the

1. strength of the joint
2. number of degrees the shafts are out of line
3. difference in the lengths of the shafts
4. combined weight of the shafts and the joint

3-68. The fixed, flexible, and Oldham couplings have a common use, which is to connect rotating shafts that are

1. perfectly aligned
2. misaligned by more than 25°
3. slightly misaligned
4. severely stressed

3-69. What advantage does a vernier-type coupling have over a sliding lug coupling?

1. Simplicity of operation
2. Strength
3. Flexibility
4. Accuracy of adjustment

3-70. Cams are generally used for all of the following purposes EXCEPT

1. transmitting power
2. changing the direction of motion from up and down to rotary
3. controlling mechanical units
4. synchronizing two or more engaging units

3-71. When the valve of figure 3E is not being lifted by the cam lobe, the cam roller is held in contact with the edge of the cam by the

1. speed of the camshaft
2. spring as it shortens
3. weight of the valve
4. spring as it lengthens

3-72. A function of the clutch in the drive mechanism of a power boat is to

1. permit changes in gear ratio
2. disconnect the engine from the propeller shaft
3. reverse the pitch of the propeller
4. reverse the direction of the engine rotation

3-73. What type of clutch has interlocking teeth?

1. Single disk
2. Cone
3. Hele-Shaw
4. Spiral claw

3-74. Either a positive clutch or a friction clutch may be used in a gear train to

1. obtain a greater mechanical advantage
2. synchronize gear speeds before the gears are meshed
3. permit interruption of power transmission through the train
4. compensate for slight angular misalignment of shafts

3-75. Magnetic and induction clutches differ mainly in the manner in which the

1. movable clutch face is actuated
2. contacting surfaces are lubricated
3. driving and driven faces are brought into contact
4. power is transmitted between the driving and driven members

ASSIGNMENT 4

Textbook
Assignment:

"Internal Combustion Engine," chapter 12, pages 12-1 through 12-23 and "Power Trains," chapter 13, pages 13-1 through 13-18.

4-1. An internal combustion engine is a machine that converts

1. heat energy to mechanical energy through the burning of a liquid fuel
2. mechanical energy to heat energy through the burning of a liquid fuel
3. mechanical energy to heat energy through the burning of a fuel-air mixture within itself
4. heat energy to mechanical energy through the burning of a fuel-air mixture within itself

4-2. All internal combustion engines rely on which of the following three things?

1. Oil, water, and air
2. Fuel, water, and ignition
3. Air, fuel, and ignition
4. Air, ignition, and water

4-3. In the operation of a gasoline engine, what event forces each piston downward?

1. Compression of fuel-air mixture
2. Intake of fuel-air mixture
3. Expansion of heated gases
4. Exhaust of waste gases

4-4. What are the four basic parts of a 1-cylinder internal combustion engine?

1. Crankshaft, piston, connecting rod, and cylinder
2. Piston, crankpin, cylinder, and crankshaft bearing
3. Crankshaft bearing, cylinder, connecting rod, and exhaust port
4. Cylinder, intake port, exhaust port, and piston

4-5. In what order do the strokes of a 4-stroke Otto-cycle engine occur during operation?

1. Compression, power, exhaust, intake
2. Compression, power, intake, exhaust
3. Intake, compression, power. exhaust
4. Intake, compression, exhaust, power

4-6. During which complete stroke of a gasoline engine is the cylinder pressure less than atmospheric pressure?

1. Compression
2. Power
3. Intake
4. Exhaust

4-7. Which of the following events occurs during a compression stroke in the 4-stroke Otto-cycle engine?

1. A partial vacuum is created
2. Waste gases are exhausted
3. Volume of air-fuel mixture decreases
4. Temperature of air-fuel mixture decreases

4-8. How are the pressure and temperature affected in an engine cylinder as the air-fuel mixture is compressed?

1. Pressure and temperature decrease
2. Pressure and temperature increase
3. Pressure decreases; temperature increases
4. Pressure increases temperature decreases

4-9. During what stroke in the operating cycle of a 4-stroke Otto-cycle engine is the greatest force exerted on the piston head?

1. Intake
2. Compression
3. Power
4. Exhaust

4-10. Which of the following events occurs during the exhaust stroke in a 4-stroke Otto-cycle engine?

1. Fuel-,air-mixture is ignited
2. Temperature and pressure of mixture increases
3. A partial vacuum is created
4. Burnt gasses are cleared from the cylinder

4-11. The basic difference between the 2-stroke-cycle and the 4-stroke-cycle diesel engine is in the

1. number of pistons
2. piston arrangement
3. number of piston strokes during a cycle of events
4. distance is piston travels during a stroke

4-12. How many crankshaft revolutions are required for each power stroke in a (a) 4-cycle engine and (b) 2-cycle engine?

1. (a) Two (b) one
2. (a) Four (b) two
3. (a) One (b) two
4. (a) Two (b)) four

4-13. Which, if any, of the following components determine(s) the position of the valves?

1. The pistons
2. The camshaft
3. The crankshaft
4. None of the above

4-14. The ignition system is timed so that the spark occurs before the piston reaches TDC on which of the following strokes?

1. Exhaust
2. Intake
3. Power
4. Compression

4-15. Which of the following engine classification methods is the most common?

1. Type of fuel used
2. Cylinder arrangement
3. Valve arrangement
4. Type of cooling used

4-16. Combustion takes place as a result of ignition by what in a (a) diesel engine and (b) gasoline engine?

1. (a) Expansion of compressed gases
   (b) a Spark
2. (a) Heat of compression
   (b) a spark
3. (a) A spark
   (b) heat of compression
4. (a) A spark
   (b) expansion of compressed gases

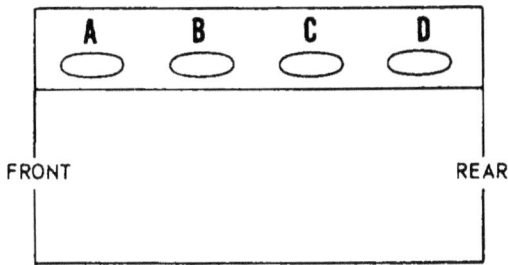

Figure 4A

IN ANSWERING QUESTION 4-17, REFER TO FIGURE 4A.

4-17. The digits in the firing order of an engine are the cylinder numbers. If the firing order of the engine of figure 4A is 1-4-2-3, in which order do the cylinders fire?

1. A, B, C, D
2. A, c, B, D
3. A, D, B, C
4. A, D, C, B

4-18. How does the camshaft actuate the intake and exhaust valves of an L-head engine?

1. By tappets from a position above the valves
2. By tappets from a position below the valves
3. By tappets, pushrods, and rocker arms from a position above the valves
4. By tappets, pushrods, and rocker arms from a position below the valves

4-19. Which of the following is NOT considered to be a stationary part of an engine?

1. The piston assembly
2. The cylinder block
3. The crankcase
4. The cylinder head

4-20. Cylinder sleeves for the blocks of gasoline and diesel engines are used for which of the following purposes?

1. To decrease the wear of the cylinder blocks
2. 10 strengthen the cylinder blocks
3. To help enclose the heat in the cylinder blocks
4. To help make a seal to contain the oil within the cylinder blocks

4-21. The curved surface of the pockets in which the valves of an L-head cylinder head function are designed for which of the following purposes?

1. To shorten the compression stroke
2. To lengthen the compression stroke
3. To decrease the turbulence of the air-fuel mixture
4. To increase the turbulence of the air-fuel mixture

4-22. Which of the following components supports and encloses the crankshaft and provides a reservoir for the lubricating oil?

1. The cylinder head
2. The exhaust manifold
3. The intake manifold
4. The crankcase

4-23. The waste products of combustion are carried from the cylinders through which of the following means?

1. The intake manifold
2. The exhaust manifold
3. The cylinder head
4. The cylinder block

4-24. Downward motion of the pistons is converted to rotary motion through the action of which of the following components?

1. The valves
2. The gear train
3. The flywheel and the vibration dampener
4. The connecting rod and the crankshaft

4-25. Which of the following parts is NOT a structural component of a piston?

1. The ring grooves
2. The lands
3. The bearings
4. The skirt

4-26. Aluminum pistons will expand more than cast-iron pistons under the same operating conditions. For this reason, they are designed with which of the following types of piston skirts?

1. Split skirts
2. Full trunk skirts only
3. Slipper skirts only
4. Full trunk and slipper skirts

4-27. Which of the following parts secure(s) the piston to the connecting rod?

1. The wrist pin
2. The split skirts
3. The piston rings
4. The ring grooves.

4-28. How do piston rings help an engine perform its work?

1. By sealing the cylinder
2. By distributing and controlling lubricating oil on the cylinder wall
3. By transferring heat from the piston to the cylinder wall
4. All of the above

4-29. The bottom ring on the piston of textbook figure 12-15 serves which of the following purposes?

1. It scrapes combustion products from piston surfaces
2. It transmits oil to the combustion rings
3. It wipes excess oil from the cylinder walls
4. It removes impurities from the oil

4-30. The end of the connecting rod that attaches to the piston must be fitted with a bearing of bronze! or similar material when the piston pin is a

1. full floating pin
2. fixed pin
3. full floating or a fixed pin
4. semifloating pin

4-31. Which of the following parts may be considered the backbone of the engine?

1. The pistons
2. The crankshaft
3. The connecting rods
4. The bearings

4-32. The vibration damper serves what purpose?

1. It balances camshaft speed with crankshaft speed
2. It reduces twisting strain on the crankshaft
3. It brakes the flywheel during engine speed reduction
4. It reduces flywheel vibration

4-33. In addition to reducing engine speed fluctuations, the flywheel often functions in which of the following ways?

1. As a power takeoff for the camshaft and a pressure surface for the clutch
2. As a pressure surface for the clutch and a starting system gear
3. As a starting system gear and a power takeoff for the fuel pump
4. As a power takeoff for the fuel pump and a timing reference for the ignition system

4-34. Which of the following parts is/are NOT included in the valve-actuating mechanism?

1. The pushrods
2. The rocker arms
3. The camshaft
4. The crankshaft

4-35. What is the function of the eccentric lobes on a camshaft?

1. To open the intake and exthaust valves at the proper times
2. To return the intake and exhaust valves to their seats
3. To add to the pressure exerted by the valve springs
4. To regulate the pressure exerted by the valve springs

4-36. Relative to engine speed, how fast does the camshaft of an 8-cylinder, 4-stroke/cycle engine turn?

1. One-eighth as fast
2. One-fourth as fast
3. One-half as fast
4. Twice as fast

4-37. Camshaft followers are the parts of the valve-actuating mechanism that contact the camshaft. Which of the following terms is another name for camshaft followers?

1. Cam lobe
2. Rocker arms
3. Valve stern
4. Valve lifters

4-38. Which of the following mechanisms keep the crankshaft and camshaft turning in the proper rotation to one another so that the valves open and close at the proper time?

1. The pushrods
2. The timing gears
3. The rocker arms
4. The valve mechanisms

4-39. By what means are the timing gears at the camshaft and crankshaft positioned so they CANNOT skip?

1. They are welded
2. They are threaded
3. They fire keyed
4. They are bolted

4-40. In the diesel engine fuel system, which of the following component replaces the carburetor?

1. The fuel injection mechanisms
2. The fuel pump
3. The fuel filter
4. All of the above

4-41. What power train part of a 4-wheel drive heavy truck is NOT part of a 2--wheel drive heavy truck?

1. The differential
2. The multiple disk clutch
3. The 4-speed transmission
4. The transfer case

4-42. What is the function of the clutch in the power train of a motor vehicle that is starting to move forward from a still position?

1. To dampen vibration in the transmission system
2. To allow the brakes to "clutch" or hold until there is enough power for the vehicle to move forward
3. To transmit power to the wheels through the dead axles
4. To allow the engine to take up the load gradually

4-43. If the spring pressure applied to the clutch driving plate is increased rapidly, what, if anything, happens to the amount of clutch slippage?

1. It increases gradually
2. It increases rapidly
3. It decreases rapidly
4. Nothing

4-44. When a truck having a 4-speed transmission is in fourth gear, the propeller shaft and the engine crankshaft rotate at a ratio of

1. 1:1
2. 1:2
3. 2:1
4. 3:2

4-45. A heavy truck with a 7:1 gear ratio in a 4-speed transmission is moving along at 6 miles per hour in low gear. The driver shifts the transmission through second and third to fourth gear. About how fast will the truck be moving in fourth gear if the driver keeps the engine turning at the same rate as it was turning in low gear?

1. 6 mph
2. 30 mph
3. 42 mph
4. 54 mph

4-46. How does the constant mesh transmission reduce noise?

1. By using spur-tooth rather than helical gears
2. By using helical rather than spur-tooth gears
3. By using main shaft meshing gears that are able to move endwise
4. By using soundproof padding around the transmission units

4-47. What is the function of the friction cone clutch in a synchromesh transmission?

1. To engage the main drive gear with the transmission main shaft
2. To engage the first-speed main shaft with the transmission main shaft
3. To equalize the speed of the driving and driven members
4. To engage the second-speed main shaft with the transmission main shaft

4-48. The synchromesh transmission shown in figure 13-10 of your textbook engages the notches at the inner ends of the bell cranks by which of the following means?

1. Shifter forks
2. A first-speed clutch
3. Poppets
4. A dog clutch

4-49. What device usually provides the means for engaging automatically the front-wheel drive on a 6-wheel drive vehicle?

1. The sprag unit
2. The power takeoff
3. The auxiliary transmission
4. The two-way clutch

4-50. In an automotive vehicle the power takeoff that supplies power to the! auxiliary accessories is attached to which of the following units of the power train?

1. The transmission
2. The auxiliary transmission
3. The transfer case
4. Each of the above

4-51. One final drive part of the truck shown in figure 13-1 of your textbook is tile

1. differential carrier
2. rear universal joint
3. propeller shaft
4. transmission

4-52. If the ring gear in a final drive has 21 teeth and the pinion has 7 teeth, the mechanism is probably part of a

1. diesel-powered shovel
2. small tractor
3. six-wheel truck
4. passenger car

4-53. What is the primary purpose of the differential in the rear axle assembly?

1. To connect each of the rear axle shafts together
2. To prevent each of the rear wheel axles from turning at a different speed
3. To boost engine power transmitted to the wheels
4. To permit both drive axles to be driven as a single unit

4-54. Through which parts of the differential is power transmitted directly to the axle shafts?

1. Differential case and side gears
2. Bevel drive pinions and side gears
3. Differential pinions and side gears
4. Differential case and bevel drive pinons

4-55. What parts usually found in conventional automotive differentials are NOT contained in the no-spin differential?

1. Ring gear and spider
2. Pinion and side gears
3. Spring retainer and side member
4. Driven clutch member and cam assembly

4-56. The rear axle housing of a certain truck helps carry the weight of the truck. Which of the following types of live axles is used in the truck?

1. Nonfloating
2. Semifloating
3. Three-quarter floating
4. Each of the above

www.ingramcontent.com/pod-product-compliance
Lightning Source LLC
Chambersburg PA
CBHW051215200326
41519CB00025B/7125